Benchmarking Water Utility Customer Relations Best Practices

Awwa Research Foundation

About the Awwa Research Foundation

The Awwa Research Foundation (AwwaRF) is a member-supported, international, nonprofit organization that sponsors research to enable water utilities, public health agencies, and other professionals to provide safe and affordable drinking water to consumers.

The Foundation's mission is to advance the science of water to improve the quality of life. To achieve this mission, the Foundation sponsors studies on all aspects of drinking water, including supply and resources, treatment, monitoring and analysis, distribution, management, and health effects. Funding for research is provided primarily by subscription payments from approximately 1,000 utilities, consulting firms, and manufacturers in North America and abroad. Additional funding comes from collaborative partnerships with other national and international organizations, allowing for resources to be leveraged, expertise to be shared, and broad-based knowledge to be developed and disseminated. Government funding serves as a third source of research dollars.

From its headquarters in Denver, Colorado, the Foundation's staff directs and supports the efforts of more than 800 volunteers who serve on the board of trustees and various committees. These volunteers represent many facets of the water industry, and contribute their expertise to select and monitor research studies that benefit the entire drinking water community.

The results of research are disseminated through a number of channels, including reports, the Web site, conferences, and periodicals.

For subscribers, the Foundation serves as a cooperative program in which water suppliers unite to pool their resources. By applying Foundation research findings, these water suppliers can save substantial costs and stay on the leading edge of drinking water science and technology. Since its inception, AwwaRF has supplied the water community with more than $300 million in applied research.

More information about the Foundation and how to become a subscriber is available on the Web at **www.awwarf.org.**

Benchmarking Water Utility Customer Relations Best Practices

Prepared by:
Roger Patrick
Competitive Advantage Consulting Ltd.
1018 Cypress Avenue, Jenkintown, PA 19046
and
Christine Kozlosky
The Ascent Group, Inc.
1584 Cowford Bridge Road, Kite, GA 31049

Sponsored by:
Awwa Research Foundation
6666 West Quincy Avenue, Denver, CO 80235-3098

Published by:

Awwa Research Foundation

American Water Works Association

IWA Publishing

DISCLAIMER

This study was funded by the Awwa Research Foundation (AwwaRF). AwwaRF assumes no responsibility for the content of the research study reported in this publication or for the opinions or statements of fact expressed in the report. The mention of trade names for commercial products does not represent or imply the approval or endorsement of AwwaRF. This report is presented solely for informational purposes.

CONTENTS

TABLES

FIGURES

FOREWORD

The Awwa Research Foundation is a nonprofit corporation that is dedicated to the implementation of a research effort to help utilities respond to regulatory requirements and traditional high-priority concerns of the industry. The research agenda is developed through a process of consultation with participants and drinking water professionals. Under the umbrella of a Strategic Research Plan, the Research Advisory Council prioritizes the suggested projects based upon current and future needs, applicability, and past work; the recommendations are forwarded to the Board of Trustees for final selection. The foundation also sponsors research projects through the unsolicited proposal process; the Collaborative Research, Research Applications, and Tailored Collaboration programs; and various joint research efforts with organizations such as the U.S. Environmental Protection Agency, the U.S. Bureau of Reclamation, and the Association of California Water Agencies.

This publication is a result of one of these sponsored studies, and it is hoped that its findings will be applied in communities throughout the world. The following report serves not only as a means of communicating the results of the water industry's centralized research program but also as a tool to enlist the further support of the nonmember utilities and individuals.

Projects are managed closely from their inception to the final report by the foundation's staff and large cadre of volunteers who willingly contribute their time and expertise. The foundation serves a planning and management function and awards contracts to other institutions such as water utilities, universities, and engineering firms. The funding for this research effort comes primarily from the Subscription Program, through which water utilities subscribe to the research program and make an annual payment proportionate to the volume of water they deliver and consultants and manufacturers subscribe based on their annual billings. The program offers a cost-effective and fair method for funding research in the public interest.

A broad spectrum of water supply issues is addressed by the foundation's research agenda: resources, treatment and operations, distribution and storage, water quality and analysis, toxicology, economics, and management. The ultimate purpose of the coordinated effort is to assist water suppliers to provide the highest possible quality of water economically and reliably. The true benefits are realized when the results are implemented at the utility level. The foundation's trustees are pleased to offer this publication as a contribution toward that end.

Walter J. Bishop
Chair, Board of Trustees
Awwa Research Foundation

Robert C. Renner, P.E.
Executive Director
Awwa Research Foundation

ACKNOWLEDGMENTS

The authors of this report gratefully acknowledge the following utilities and individuals for their participation and support during this project:

Beaufort-Jasper Water and Sewer Authority
City of Cleveland Division of Water
Greater Cincinnati Water Works
Denver Water
Las Vegas Valley Water District
Long Beach Water Department
The City of Los Angeles Department of Water & Power
Louisville Water Company
Passaic Valley Water Commission
City of Phoenix Water Services Department
PPL Electric Utilities
Sarasota County Environmental Services
Tucson Water Department
Washington Suburban Sanitary Commission
Yarra Valley Water

In addition, the authors wish to acknowledge the advice and support of the Awwa Research Foundation project manager, India Williams, and the Project Advisory Committee:

Rebecca Lamoreaux, East Bay Municipal Utility District
Nora Freeman, Louisville Water Company
Lyn Faas, Seattle Public Utilities
Tony Gipe, Sarasota County Environmental Services

EXECUTIVE SUMMARY

BACKGROUND AND OBJECTIVES

Customer satisfaction is important to water utilities to minimize customer complaints and associated costs, maintain customer goodwill, and increase customer support for utility improvement initiatives. A good way to improve customer satisfaction is to optimize utility processes that directly affect customers.

This project had three primary objectives, all with the ultimate goal of improving customer satisfaction and utility efficiency:

1. To identify customer relations best practices from other relevant organizations;
2. To identify metrics for both internal performance tracking and external comparison;
3. To develop tools that will enable water utilities to improve customer relations.

PROJECT SCOPE AND DELIVERABLES

Many areas of water utility operations impact customers either directly or indirectly. However, this project's scope relates only to the "classic" customer service functions: call center, meter reading, billing, payment processing, credit and collections, and customer-related field and meter services.

The major deliverables from this project are this report and the project web site, www.wisebenchmarks.org. Access to the web site requires a User ID and password, which can be obtained by following instructions found on the home page.

APPROACH

This project comprised five tasks:

1. Identify Customer Relations Functions of Focus;
2. Research Relevant Customer Relations Metrics and Practices;
3. Identify Customer Relations Best Practices;
4. Develop Web Toolkit and Report;
5. Communicate Findings.

Metrics and practices data were gathered from 15 participating utilities, for the functions of interest, using on-line surveys and interviews. These surveys are now on the web site for ongoing use by water utilities. Detailed definitions are provided on the site to assist submitting utilities to provide data in a consistent manner. When a utility submits its data, it receives instant feedback on its metrics against database averages.

A generic customer satisfaction survey was also designed and piloted in a similar way to the metrics and practices surveys. This survey had a return rate of 10-20%, which reflects its simplicity. This survey is available on the site to allow water utilities to compare their level of customer satisfaction to others in a consistent way.

High-performing organizations were identified based on their metrics and/or a reputation for superior customer service (such as winning recognized customer satisfaction awards). Descriptions of leading organizations' practices were examined for common themes. The practices of these customer service leaders forms a key part of each results chapter of this report. Case studies were also developed on many of the best performing organizations, as well as "mini case studies" focused on best practices in specific functions.

A comprehensive web-based toolkit was developed to gather data, facilitate metric and practice comparisons, allow customized best practice searches, allow multi-year performance tracking by individual utilities, and facilitate water utility networking. The web site was fine-tuned using feedback from the participating utilities, the PAC and AwwaRF. The web-based toolkit contains:

- All the metrics and practices surveys,
- The customer satisfaction survey,
- Benchmark data (on an average-only basis except for the submitting utility),
- Case studies, searchable by key word,
- Best practices information,
- Contextual information.

RESULTS

Overall, the findings of this project indicate that there is no reason that water utilities cannot provide high levels of service and do it efficiently. Many water utilities already fall into the best performer category in one or more functions of interest. In one case, a water utility was a best performer in all functions but one.

The key findings on each function appear in Chapters 3, 4, 5 and 6. Each of these chapters is structured into three main sections: an introductory section that explains the characteristics of the submitting organizations; a section that presents and analyzes the key metrics and best practices; and a list of practices that water utilities may find useful, even if not associated with best practice performers.

Common approaches adopted by best performers include:

- Performance measurement. Best performers use detailed measures of employee and group performance in both service and efficiency of delivering that service. Best performers also link these performance measures to employee rewards. Another obvious advantage of having detailed measures tracked regularly is the support that this provides to justify improvement projects, and the mechanism to measure the benefits of such projects.

- Treating the employees who serve customers as well as the customers are treated. The details vary between organizations, but include first class supervision, mentoring and coaching, and rewards for superior customer service.

- Using technology aggressively to better serve customers, including AMR, the Internet and IVR

- Cross training of employees to cope with changes in work demands

- Outsourcing wherever this makes sense from a service and efficiency viewpoint

The key metrics used for analysis of best performers appear below.

	Service Level Measures	**Unit Costs**	**Staffing/Productivity Measures**
Call Center Operations	% abandoned, % availability, service level conformance, 1st Call Resolution Rate	$ per call handled	Calls per FTE
Meter Reading	Errors, skips	$ per meter read	Meters read per FTE
Customer Billing	Customer-initiated rebills	$ per account processed	Accounts processed per FTE
Remittance Processing	Errors, funds availability	$ per payment processed	Payments processed per FTE
Credit & Collections	Net write offs (% of total revenue), % arrears (30,60,90), cost of credit	$ per delinquent account handled	Delinquent accounts handled per FTE
Business Office Operations	Wait time	$ per walk-in transaction	Transactions per FTE
Meter Maintenance	Inaccurate meters	$ per meter tested, installed, replaced, removed	Meter sets/removals/replacements per FTE and tests per FTE
Meter-related Field Services	% orders worked by due date	$ per order worked	Orders per FTE
OVERALL CUSTOMER SERVICE PERFORMANCE	Customer Satisfaction Rating	Cost per customer	Customers per CS Employee

Results Summary for Call Center

Best performers are usually using some or all of these approaches:

- Emphasizing first call resolution and using this measure to drive improvement at all levels. These organizations also put systems in place, such as real time AMR, to help agents to get the job done the first time

- Monitoring call quality against customer values and expectations, often with specialist QA personnel

- Using an IVR system that is customer-friendly. They conduct extensive customer research in conjunction with any IVR design changes or additions, train agents to promote IVR services, selectively force callers through the IVR, and monitor IVR call quality as much as calls handled by agents

- Matching agents to demand via the use of part time agents to supplement full time staff, linking multiple call centers as one virtual call center, and pooling CSRs with front office and back office staff to gain flexibility

- Conducting aptitude testing and training of CSRs. Leaders place a great deal of emphasis on having the right service personnel

- Providing feedback and rewards to CSRs on an individual, not just a group, basis

- Having a dedicated business customer call center, and providing as many services as possible for business customers via the Internet

Results Summary for Meter Reading

Best performers are usually using some or all of these approaches:

- Reducing costs of manual reads through contract negotiations, rerouting, more sophisticated hand-held equipment and meters, productivity improvement, and lowering overhead.

- Contract meter reading or sharing meter reading with other utilities (water and/or energy) to reduce overhead, tackle seasonal peaks, or as a strategy to transition to AMR

- Automated meter reading – if not large-scale implementation, often to address high read cost meters, unsafe meter locations, and high-turnover premises

Many utilities are considering expanding AMR usage, and there are newer technologies that offer much more functionality than earlier drive by systems, making it easier to justify full AMR.

Most utilities are not reading meters for others, despite the prospect of lower cost. Those that use this approach find it successful overall.

Fears about contract meter reading are not borne out in practice; in fact the setting of performance targets as contract terms improves reading accuracy.

Results Summary for Billing, Remittance Processing, Credit and Collections

Best performers are usually using some or all of these approaches:

- Providing a multitude of payment options, including self-service payment options over the Internet and via IVR
- Either eliminating over the counter service, since there are now so many other ways to pay, and having agents such as banks provide in-person payment service, or pooling front office staff with call center agents
- Converting as many payments to electronic formats before processing them
- Having an aggressive, multi-faceted credit and collections policy, while at the same time being realistic and helpful to customers who are having a genuine difficulty in paying
- Use credit information to segment and rank collection work
- Have a consistent credit and collection policy and do what they say they'll do
- Network with credit bureaus and other utilities to identify problem customers, recover unpaid finals

Results Summary for Meter Maintenance/Meter Related Field Services

Best performers are usually using some or all of these approaches:

- Mobile dispatch and the use of portable computing power in the field
- Using a blend of contractors and staff, and cross-training, to optimize resources to varying workloads
- Prioritizing customer-request orders over internally generated orders
- Deploying cellular phones to improve communication between field and office
- Letting field employees start work day from home
- Emphasizing safety and enhancing safety training

CHAPTER 1
INTRODUCTION

PROJECT BACKGROUND

Customer satisfaction is important to water utilities to minimize customer complaints and associated costs, maintain customer goodwill, and increase customer support for utility improvement initiatives. A good way to improve customer satisfaction is to optimize utility processes that directly affect customers.

Furthermore, almost every organization realizes that good customer relations are a key to survival and success. One needs only to look at examples, good and bad, to recognize the importance of maintaining the support of the customer base. For example, a decade ago Continental Airlines ranked last amongst large airlines in customer relations and faced bankruptcy. As part of the airline's turnaround, Continental initiated a concerted effort to improve its image with customers, and for the past few years Continental has ranked first amongst its peers in customer satisfaction by JD Power and Associates. There is little doubt that such a transformation was essential to the airline's survival. As another example, Dell Computer has become America's most successful PC manufacturer largely because it developed processes that are highly responsive to customers' needs.

Even when business is flowing smoothly, water utilities need the cooperation of their customers—for example, in complying with water restrictions during droughts and in supporting rate increases that fund infrastructure upgrades. These needs will only increase in the future.

Finally, water utilities are also facing well-known funding challenges. In the face of competing demands from other areas of the utility, customer service professionals need to be able to assess areas in need of improvement, and to better target programs to improve customer satisfaction.

PROJECT OBJECTIVES

This project's intent was to help water utilities improve their customer relations by improving their understanding of practices that impact customer relations, and develop performance measurements and targets that are consistent with their customer service goals and with customer service norms. The project also aimed to identify best practices and measures associated with good customer service and efficient delivery of that service.

In summary, this project had three primary objectives, all with the ultimate goal of improving customer satisfaction and utility efficiency:

1. To identify customer relations best practices from other relevant organizations;
2. To identify metrics for both internal performance tracking and external comparison;
3. To develop tools that will enable water utilities to improve customer relations.

PROJECT SCOPE

Many areas of water utility operations impact customers either directly or indirectly. However, this project's scope relates to the "classic" customer service functions: call center, meter reading, billing, payment processing, credit and collections, and customer-related field and meter services.

With respect to the first objective (best practices), experience indicates that a wide range of industries can contribute valuable experiences in the field of customer relations. In fact, it was felt likely that many leading customer relations practices would be found outside the water industry, especially by examining customer service leaders in competitive industries. Energy utilities were considered an especially relevant industry to examine, since most customer service processes are similar to those of water utilities and many are now facing competition for customers. Such market forces are shifting energy utilities' mindset, systems, and customer service approaches, and their experiences were expected to provide especially valuable insights for water utilities.

However, we did not neglect water utilities in the search for best practices. In fact, we already knew of many good practices within our participating utility group, and set out to draw upon their experiences as well as those of other leading water utilities.

With regard to the second objective (metrics), the need was recognized for a set of detailed metrics for customer service functions that can be used throughout the water industry. While individual utilities do not need to use the same metrics as others for internal purposes, only the use of a consistent set of metrics will allow water utilities to compare performance with each other or with other types of organizations. So it was decided to define a standardized set of customer service metrics, drawing upon recognized cross-industry standards, to provide valid water utility and cross-industry comparisons.

To achieve the third objective (improvement tools) it was decided to develop a web site with a variety of customer relations measurement and improvement tools. These tools provide ongoing assistance to water utilities, as the web site will outlive the project. The tools include self-service metric comparison both within the water industry and to other industries, practices to consider, searchable case studies from a variety of industries, and a standardized customer satisfaction survey that allows water utilities to compare their level of customer satisfaction to other water utilities across the nation.

PROJECT DELIVERABLES

The major deliverables from this project are this report and the project web site, www.wisebenchmarks.org. Access to the web site requires a User ID and password, which can be obtained by following instructions found on the home page.

CHAPTER 2
APPROACH

SUMMARY

Every organization operates within a unique set of circumstances, and therefore there is no universal set of best practices that will suit them all. This project reflected this fact by documenting a variety of practices shown to be effective in promoting good customer relations, enabling water utilities to select the practices most beneficial to them. Many of these practices are highlighted in this report, while they and more can be found on the project web site (www.wisebenchmarks.org). This self-service site is continuously being added to by water utilities, and thus provides a useful tool and growing database of customer service metrics and practices.

This project comprised five tasks:

1. Identify Customer Relations Functions of Focus;

2. Research Relevant Customer Relations Metrics, Practices and Customer Satisfaction;

3. Identify Customer Relations Best Practices;

4. Develop Web Toolkit and Report;

5. Communicate Findings.

To identify high-performing utilities and other organizations, a multi-faceted performance comparison framework was developed. This framework included metrics and practices for each customer relations process of focus. These metrics and practices covered a balance of service, productivity, and cost factors.

The metrics were chosen to maximize the ability of utilities to compare their customer relations performance to other organizations (utilities and other types of organizations), and to facilitate the identification of high-performing organizations.

Companies exhibiting above average performance in both service levels and efficiency in a particular function were designated as best performers for that function. All of the participating utilities, and many other organizations, that fell into the best performer categories were then investigated to learn more about their winning strategies and practices. A summary of commonly used approaches to achieving high efficiency and customer satisfaction appears in the Executive Summary of this report.

A key element of the approach was the development of a web-based toolkit designed to assist utilities to identify customer relations improvement opportunities, assess customer relations performance and satisfaction, measure performance and customer satisfaction over time and against others, and discover ideas for improvement that have been successfully implemented by leaders in customer relations.

All of the metrics, practices and customer satisfaction surveys used in this project are available on the web site for other water utilities to use.

Study participants are listed in the Acknowledgements section of this report. Industry segments represented in the web site database are shown Figure 2.1 below.

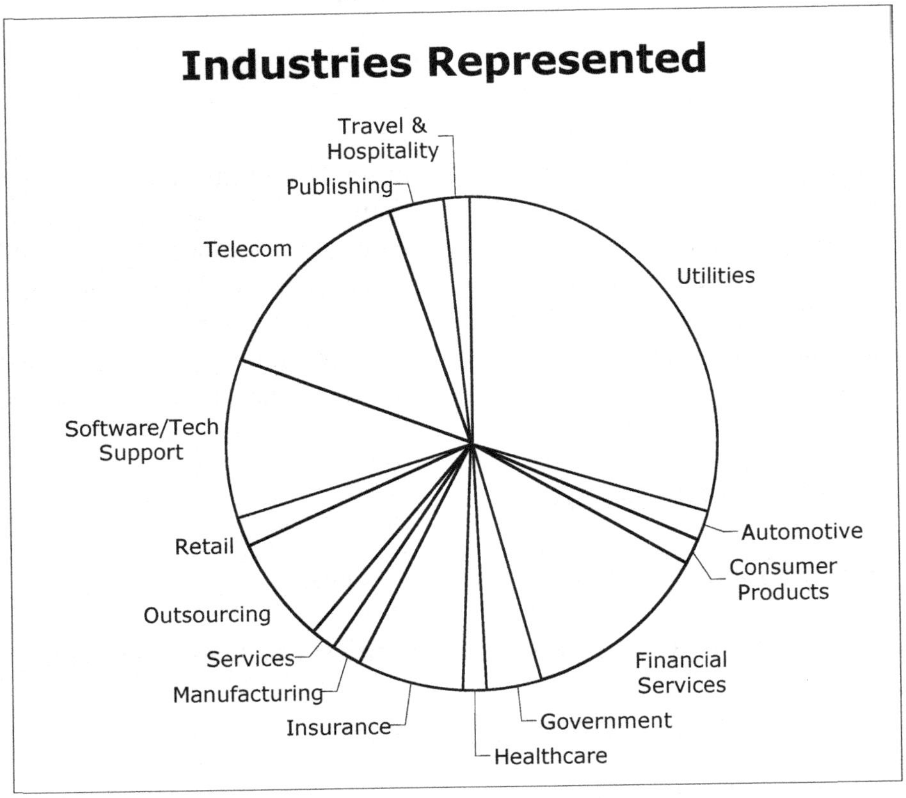

Figure 2.1 Industry segments represented

TASK 1: IDENTIFY CUSTOMER RELATIONS FUNCTIONS OF FOCUS

The focus of the benchmark and best practice research was on the "core" customer service functions, namely:

- Customer contact functions: call center and front office,
- Revenue cycle functions: meter reading, billing, remittance processing, and credit and collections,
- Meter maintenance and meter-related field services

TASK 2: RESEARCH RELEVANT CUSTOMER RELATIONS METRICS, PRACTICES AND CUSTOMER SATISFACTION

Data were gathered using several surveys which were all pilot tested by the participating utilities. Feedback from the participants on the surveys was taken into account prior to their finalization. These surveys are now on the web site for ongoing use by water utilities.

Participating utilities used the web site to submit their metrics data and practices information for each of the customer relations functions identified in Task 1. As well as accepting the data, the site can give instantaneous feedback on over 100 relevant metrics, by comparing an individual utility's metrics to database averages for each of these metrics.

Detailed definitions are provided on the site, to assist submitting utilities to provide data in a consistent manner. Also, when a utility submits its data, the feedback it receives against database averages allows a self-check on data accuracy.

To designate best performers, a core set of metrics was defined, as shown in Table 2.1 below.

Table 2.1 Benchmark Metric Framework

	Service Level Measures	**Unit Costs**	**Staffing/Productivity Measures**
Call Center Operations	% abandoned, % availability, service level conformance, 1st Call Resolution Rate	$ per call handled	Calls per FTE
Meter Reading	Errors, skips	$ per meter read	Meters read per FTE
Customer Billing	Customer-initiated rebills	$ per account processed	Accounts processed per FTE
Remittance Processing	Errors, funds availability	$ per payment processed	Payments processed per FTE
Credit & Collections	Net write offs (% of total revenue), % arrears (30,60,90), cost of credit	$ per delinquent account handled	Delinquent accounts handled per FTE
Business Office Operations	Wait time	$ per walk-in transaction	Transactions per FTE
Meter Maintenance	Inaccurate meters	$ per meter tested, installed, replaced, removed	Meter sets/removals/replacements per FTE and tests per FTE
Meter-related Field Services	% orders worked by due date	$ per order worked	Orders per FTE
OVERALL CUSTOMER SERVICE PERFORMANCE	Customer Satisfaction Rating	Cost per customer	Customers per CS Employee

These core metrics were chosen as they:

- Reflect a balance between service and efficiency,
- Are typical of metrics used widely by utilities of all types, and other types of organizations,
- Are based on data commonly available from water utilities,
- Match existing data available to the project team from other organizations,
- Are likely to be available going forward

With the assistance of the participating utilities, many of whom provided their existing customer satisfaction surveys, a generic customer satisfaction survey was designed. To test this survey, four of the participants mailed it to at least 4,000 customers. The survey return rate in all cases was over 10%, which is good for this type of survey and reflects its simplicity. These utilities then entered their aggregated survey results on the web site.

This customer satisfaction survey allows any water utility to compare its customer satisfaction to other water utilities. It also allows utilities to track their levels of customer satisfaction over time by repeating the survey periodically, and also to compare satisfaction between subsets of their customers (eg different townships or groups within their service territory).

As more water utilities use the survey, the database will grow and hopefully become a reference tool for measuring customer satisfaction on a consistent basis across the industry.

TASK 3: IDENTIFY CUSTOMER RELATIONS BEST PRACTICES

High-performing organizations were identified based on their metrics and/or a reputation for superior customer service (such as winning recognized customer satisfaction awards). Examples of organizations identified by reputation are PPL Electric Utilities and LG&E Energy, two of the nations highest ranking electric utilities for customer service according to the June 2004 JD Power customer satisfaction survey. PPL, ranked highest in the Eastern Region for the fourth consecutive year, was a participating utility in this project, and LG&E ranked highest in the Midwestern Region for the fifth time in six years. Both utilities were visited during the project and provided case studies.

Descriptions of practices submitted on the web site by the identified leading organizations were examined for similarities among their practices.

Case studies were developed on many of the best performing organizations, focused on best practices in specific functions. All of the case study providers in this report have achieved excellent customer satisfaction as measured by metric analysis conducted during this project or independent external surveys (JD Power, ACSI or other).

The case studies were prepared by interviewing customer service managers from the best performers, by conducting site visits, or by reviewing submissions on the web site by the organizations concerned. These case studies appear within the relevant results chapters to illustrate practices that lead to superior performance. Also, 10 of the best performers overall are covered in comprehensive case studies in Chapter 8 of this report.

TASK 4: DEVELOP WEB TOOLKIT AND REPORT

A comprehensive web-based toolkit was developed to facilitate metric and practice comparisons, allow customized best practice searches, allow multi-year data tracking by individual utilities, and facilitate water utility networking after completion of this project. The web site was fine-tuned using feedback from the participating utilities, the PAC and AwwaRF.

The web site was also used by the participating water utilities to obtain performance feedback, and it can be used on an ongoing basis to inform and inspire positive change.

The web-based toolkit contains:

- All the metrics and practices surveys,
- The customer satisfaction survey,
- Benchmark data (on an average basis, except for the submitting utility),
- Case studies, searchable by key word,
- Best practices information,
- Contextual information.

The site also contains up-to-date information on industry vendors, recent relevant press announcements and a compendium of useful customer service articles.

As metrics, practices, and customer survey information is submitted to the site, the database is automatically updated to reflect the new information. Utilities that have access to the site are able to compare themselves to aggregate results covering over 100 customer service metrics, and search for relevant best practices.

The web site will be maintained by the researchers until the end of 2008, to retain access for water utilities to the information and tools.

This report highlights key findings from the project. Chapters 3 through 6 of this report each cover a function of interest, and are structured into three main sections:

- An introductory section that explains the characteristics of the submitting organizations,
- A section that presents the key metrics and practices displayed by the best performers,
- A list of useful practices to consider, even if not associated with best performers.

TASK 5: COMMUNICATE FINDINGS

As well as producing this report, the project researchers made presentations on this project at the AWWA Customer Service Conferences and the AWWA/WEF Joint Management Conferences in 2004 and 2005.

CHAPTER 3
CALL CENTER OPERATIONS

This chapter's focus is on call center metrics and practices, with an emphasis on key metrics and practices used by best performing organizations, and practical recommendations to water utilities.

TYPICAL PRACTICES

First call resolution seems to be one of the best predictors of customer satisfaction, in other words, the higher the first call resolution, the higher the customer satisfaction. But many companies have found it difficult to define first call resolution, much less measure it consistently. This is due to a number of factors such as different views between customers and service representatives as to whether a call resolved the problem (internal first call resolution scores are typically higher than those based on customer surveys), or having supervisors assess calls as to whether the customer's issue was resolved.

Call quality monitoring refers to the process of listening to or observing an agent's phone conversations or other multi-media contacts with customers. Call quality monitoring is one of the most effective methods for improving the level of service provided to customers. Not only can it improve the customer experience, it can also improve overall call center performance, reduce callbacks, identify training or process improvement opportunities, and facilitate employee development.

In addition, many companies have created specialized Quality Assurance (QA) groups to monitor agent performance. The technology and QA groups have made it possible to conduct more monitoring sessions per month per employee than were possible in the past.

The majority of participants do not maintain a "24 x 7" call center— only 43 percent are open 24 hours a day, 365 days a year (Figure 3.1).

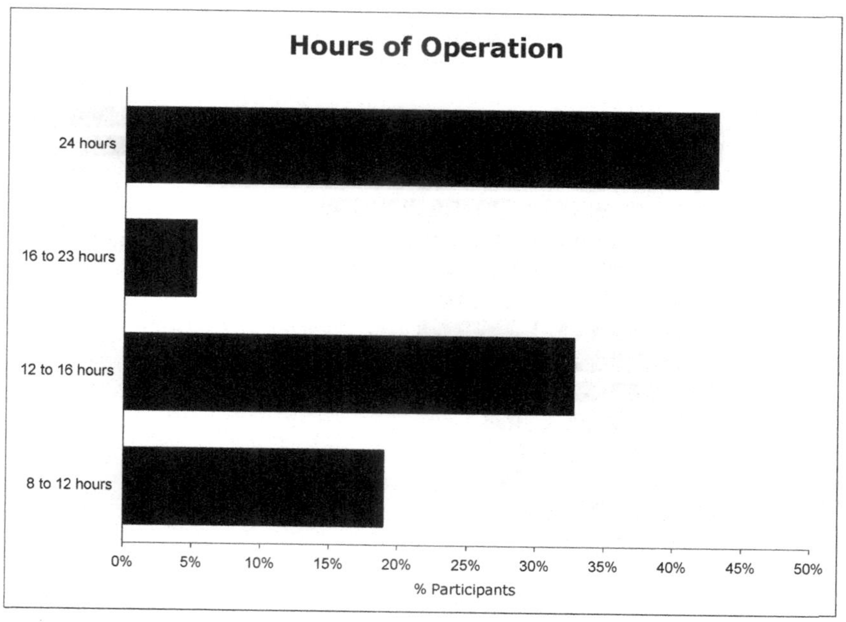

Figure 3.1 Hours of operation.

Participants handle anywhere from 10,000 calls per year to as many as 54 million. Centers are staffed by as few as six representatives to as many as 5,200, averaging about 100 agents (FTEs—full time equivalents) per center. Close to 20 percent of participants indicated that their call center representatives are union-represented or partially union-represented.

Nearly three-quarters of participants reported outbound call volumes of less than 10 percent (i.e. participants overwhelmingly run customer service call centers, not telemarketing businesses). Participants average 23 percent of calls handled through Interactive Voice Response (IVR) technology (Figure 3.2).

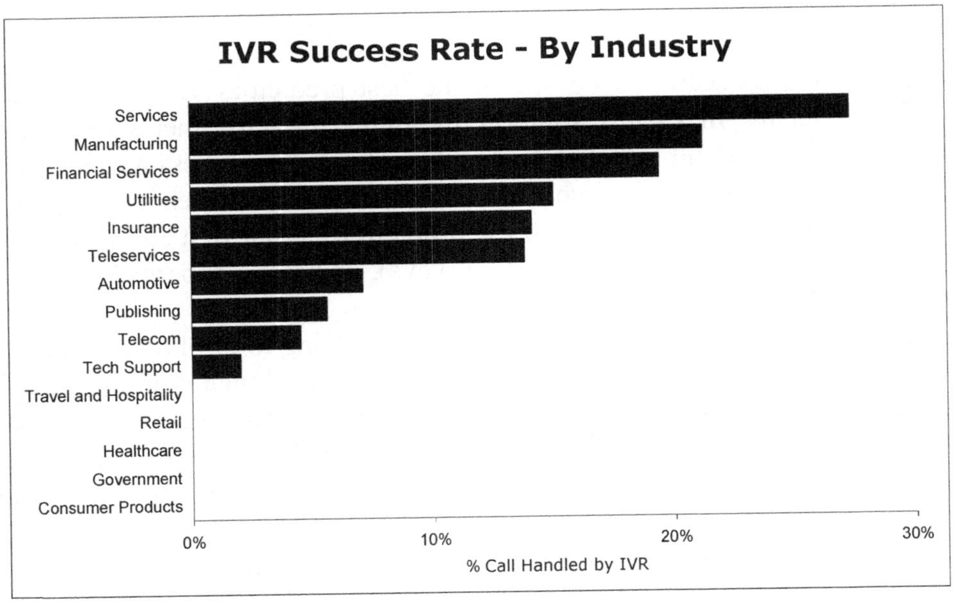

Figure 3.2 IVR success rate – by industry.

The average span of control of call center supervisors appears in Figure 3.3 below.

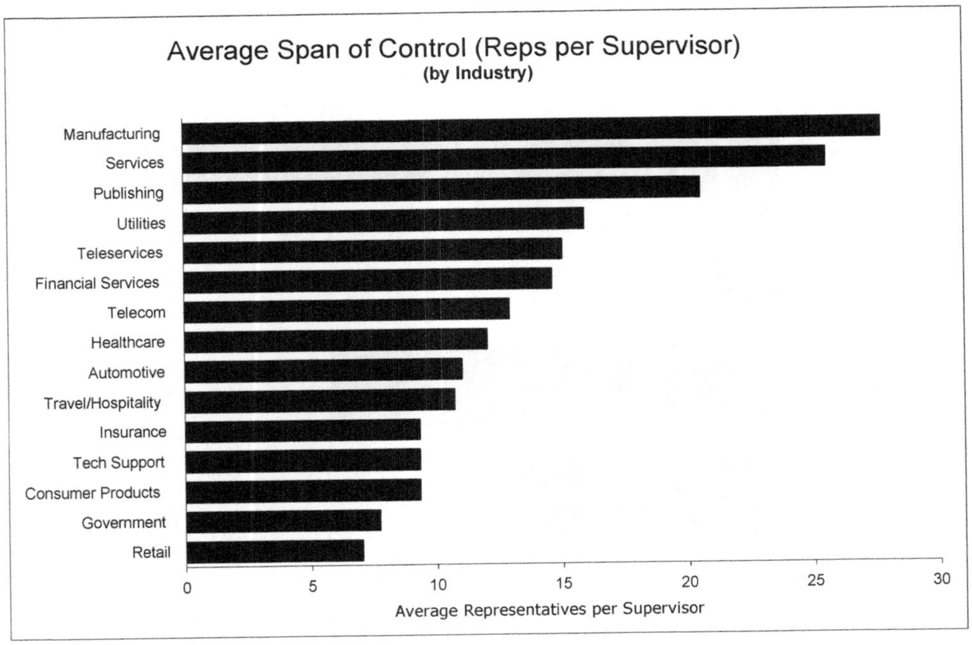

Figure 3.3 Average span of control (reps per supervisor) by industry

All participants monitor phone calls (Figure 3.4). About half also monitor email correspondence. Only 25 percent monitor data entry or keystrokes. And fewer participants monitor letters, faxes, or chat.

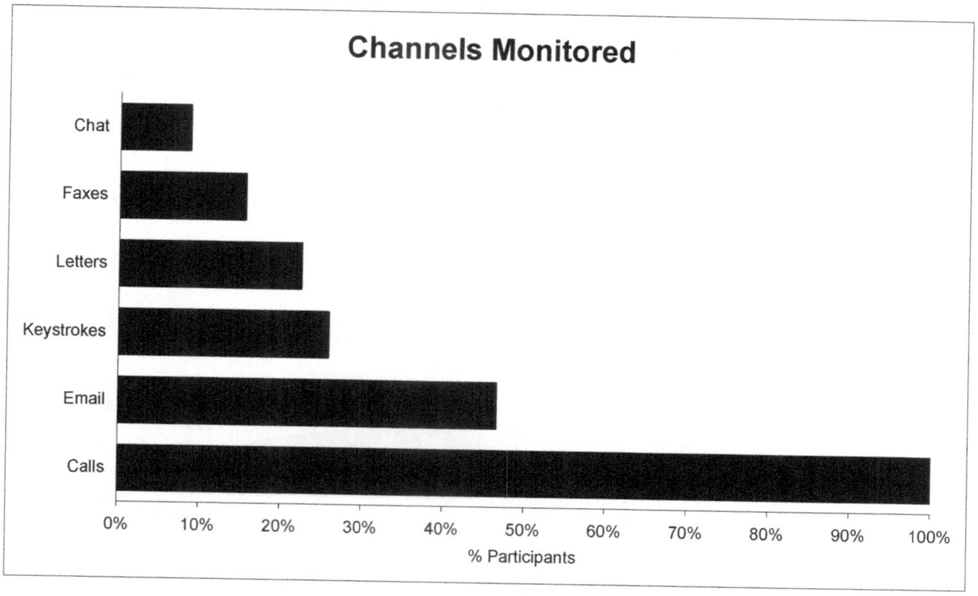

Figure 3.4 Channels monitored.

Participants vary widely on the number of calls monitored per month per representative (Figure 3.5). By industry, Healthcare (20 calls per month) and Retail (20 calls per month) report the highest number of calls monitored per employee per month. As a group, participants average about 8 monitoring sessions per month per employee.

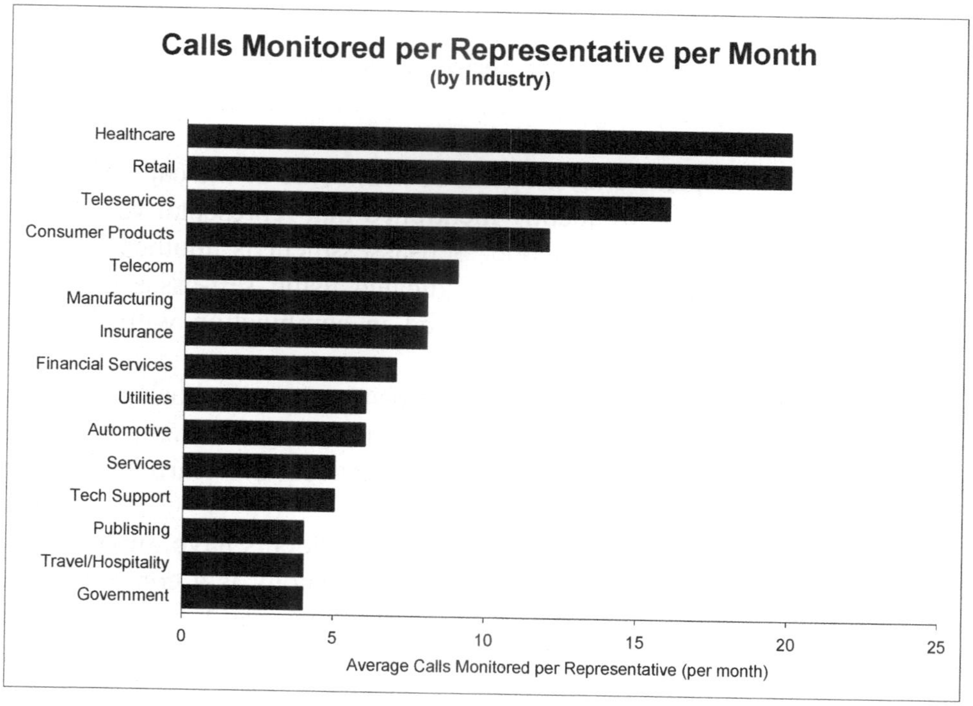

Figure 3.5 Calls monitored per representative per month, by industry.

Sixty-five percent of respondents have a Quality Assurance (QA) group of some form in place. The majority of participants rely on a combination of supervisory and QA effort to monitor call quality.

Most participants (75 percent) use a combination of side-by-side monitoring and silent (remote) monitoring through live or recorded calls (Figure 3.6). Silent or remote monitoring fosters a more natural call-handling environment, as the representative is usually unaware that monitoring is taking place. On the other hand, side-by-side monitoring evaluates representatives in "real time" with full knowledge of the employee.

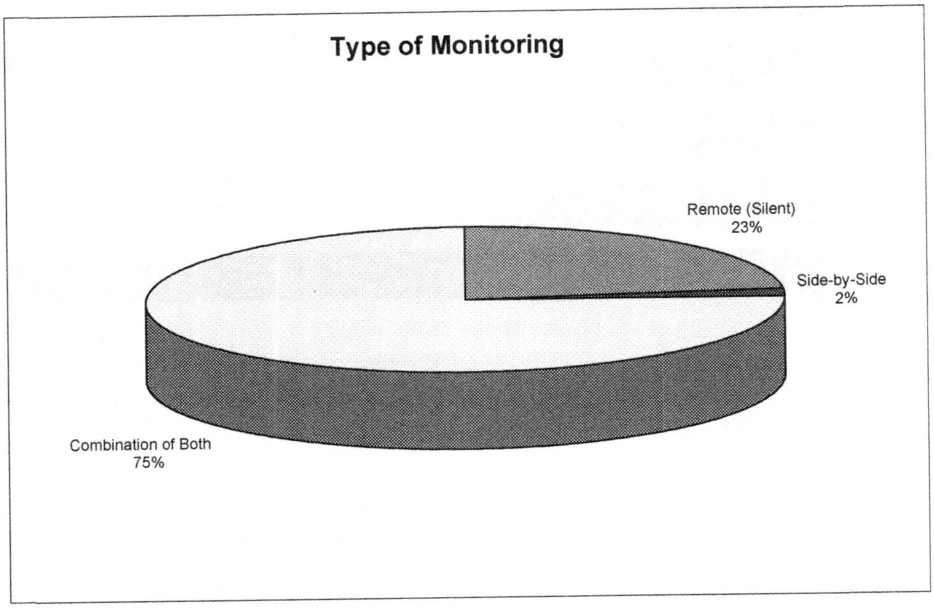

Figure 3.6 Type of monitoring.

Most participants include monitoring performance as a component in the annual review process for representatives. Only twelve percent reported that monitoring performance is not a part of the employee review process.

Three-fourths of participants enable agents to hear some of the calls monitored as part of the feedback process. Respondents frequently mentioned that representatives respond positively to listening how they sound on the telephone. Self-assessment is another powerful employee development technique. But only five percent indicated using self-assessment as part of the feedback process. The self-scoring can be reviewed in combination with the supervisor's or monitor's observations.

Twenty-five percent of respondents reported using peer monitoring to enhance training and quality efforts. Peer monitoring has many benefits, including more employee involvement in the monitoring process, opportunities to share best practices and learn from each other, and education of the front-line on the monitoring process.

As shown in Figure 3.7, sixty-four percent of respondents have implemented automated call recording and monitoring technology. The latest technology is a combination of hardware and software, records a random selection of calls, so that they can be reviewed later. Calls are stored as digital files, facilitating recall and review as needed.

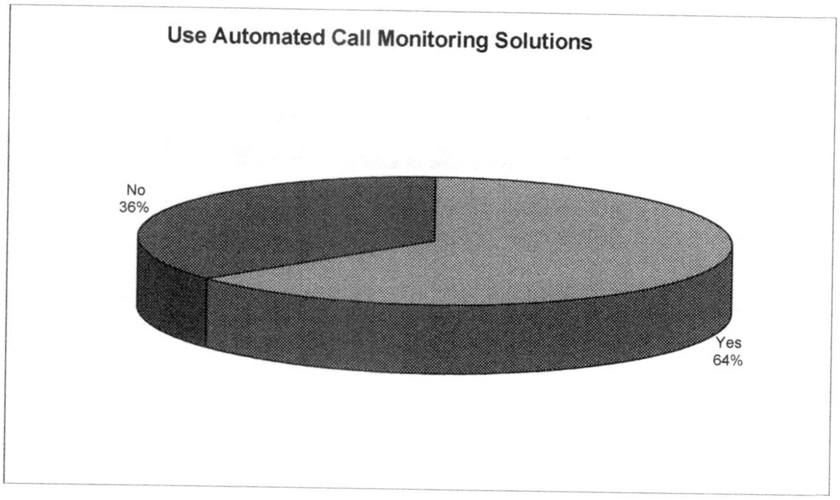

Figure 3.7 Use automated call-monitoring solutions

To ensure the objectivity and consistency of the scoring and feedback of call quality, nearly 80 percent of participants conduct regular calibration sessions (Figure 3.8). Most respondents reported that a combination of supervisors and QA team members attend typical calibration sessions, where they listen to and rank a selection of calls for quality, and compare their rankings to improve consistency of call assessment. About 30 percent include management in the calibration process. Twenty-two percent include members from the training team. Clients and customers are also sometimes included.

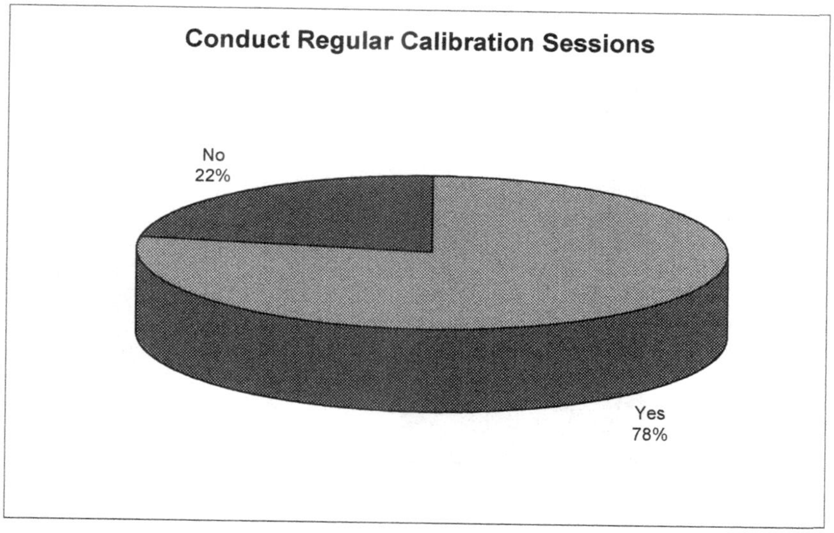

Figure 3.8 Conduct regular calibration sessions

Nearly forty percent judge the performance of coaches based upon their customer service team's performance, as shown in Figure 3.9. Others rely on more traditional management observation and evaluation. Several actively solicit feedback from representatives to gauge coaching performance, through employee opinion surveys, focus groups, one-on-one discussions and "360 degree" evaluations (a technique that provides feedback to an employee from supervisors, peers, customers and direct reports).

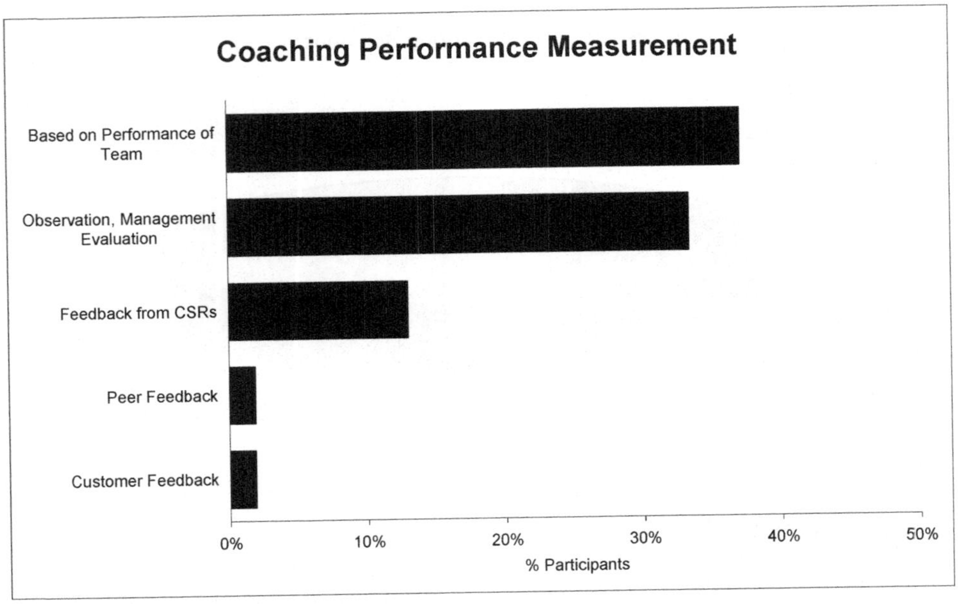

Figure 3.9 Coaching performance measurement

Most companies rely on internal training department resources to provide formal coaching or leadership training to employees and supervisors. Very few respondents use coaching certification or outside coach training programs (Figure 3.10).

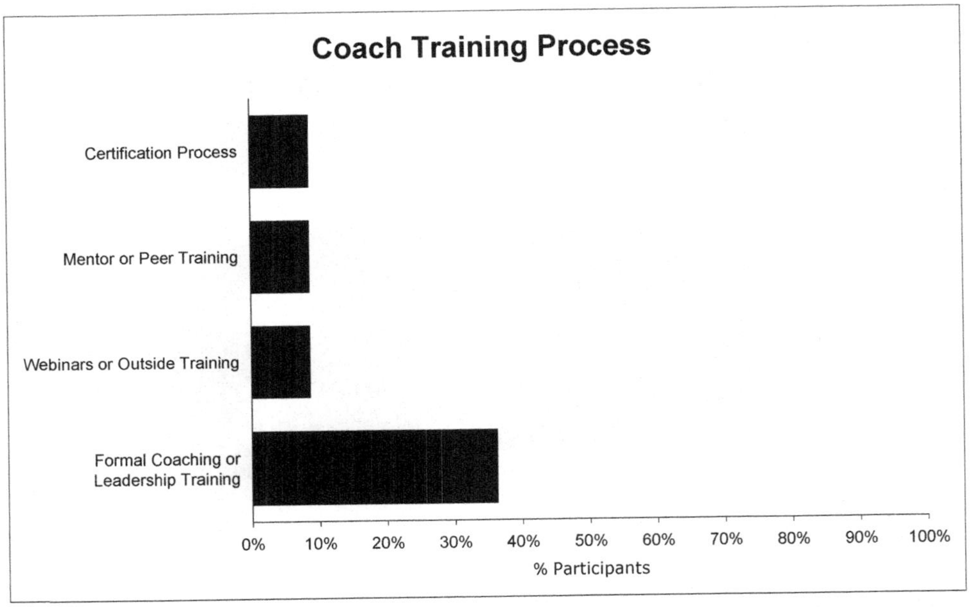

Figure 3.10 Coach training process

Almost all respondents reported the ability to routinely measure First Call Resolution (FCR), as shown in Figure 3.11. The majority of participants rely on a single measure to gauge FCR. However, sixteen percent reported measuring FCR through a series of measures.

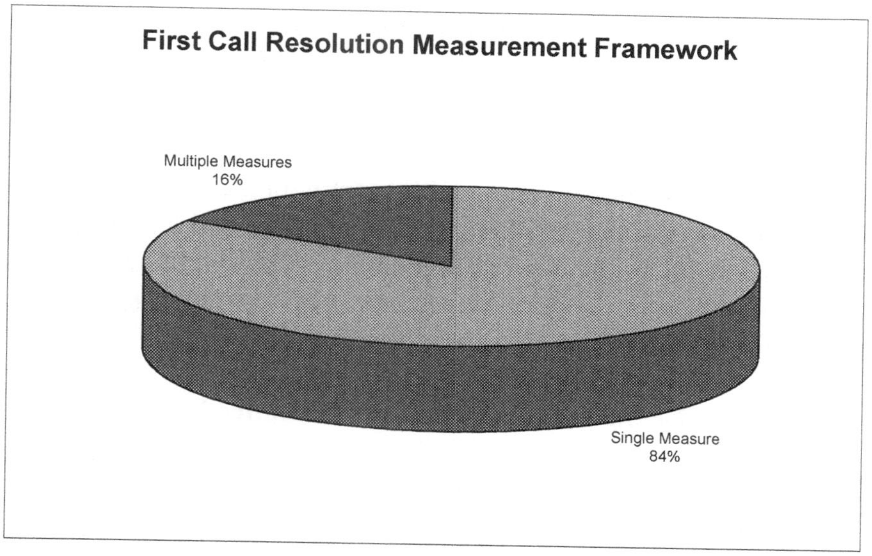

Figure 3.11 First call resolution measurement framework

Within our participant group, there are 4 primary ways of measuring first call resolution, i.e. properly addressing a customer's need on the first call (Figure 3.12), including:

- Call statistic calculations (internal measure)
- Agent-driven call logging or tick sheet (internal measure)
- Call quality monitoring (internal determination)
- Customer satisfaction surveys (external measure)

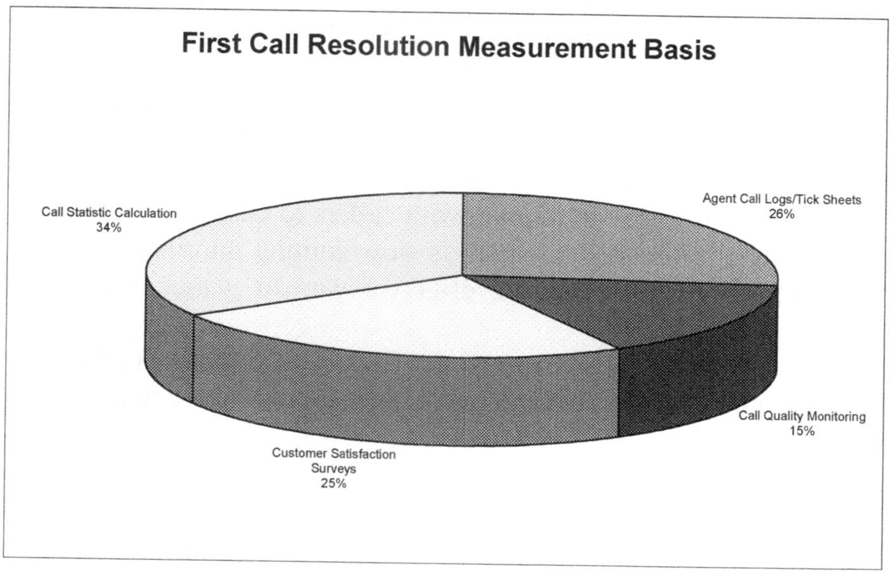

Figure 3.12 First call resolution measurement basis

Nearly two-thirds of respondents indicated using first call resolution as a component of overall call center performance. Of those that didn't, all expressed an interest in deriving a reliable, consistent measure of first call resolution.

First Call Resolution (FCR) performance varied from industry to industry, as shown in Table 3.1 below. FCR performance (average FCR rate) is tied closely to the type of calls being handled in the call center. Consumer products, manufacturing, and technical support call centers generally have lower FCR due to the nature of incoming calls—all product issues, complaints, or problems. Many of these calls require assistance from product research, parts, or other organizations groups for resolution.

Table 3.1 First call resolution performance by industry

	Average FCR Rate	Average FCR Goal	Avg Years Measuring FCR
Consumer Prod	67%	87%	4
Financial Services	78%	83%	2
Govt	74%	81%	2
Healthcare	48%	75%	3
Insurance	84%	85%	4
Manufacturing	35%		2
Outsourcing	82%	88%	6
Publishing	53%	66%	1
Retail	70%	85%	2
Services	94%	95%	8
Tech Support	70%	73%	4
Telecom	74%	77%	2
Utilities	78%	81%	2

Interactive Voice Response (IVR) is a telecommunications system that accepts a combination of voice and telephone touch-tone keypad input and provides appropriate responses in the form of voice, fax, callback, e-mail and perhaps other media.

In the technology's infancy, Auto Attendants (AA) and Voice Response Units (VRU) provided menu options and scripting tools to direct callers to certain queues or to provide information to callers, with minimal interaction. IVR brought integration with enterprise information systems and the ability to interact with callers to customize questions and responses. Automated Speech Recognition (ASR), which is slow coming into its own, adds conversational interaction with callers as well as a more effective way to gather input for interaction and customization.

In short, IVR technology offers companies more cost effective call management through call segmentation, automated call handling, and informational messaging. IVR offers customers 24-hour services and privacy.

Companies hoping to reap the benefits of IVR must understand that the customer experience is everything, even with ASR. While callers are generally happier with voice-driven systems, a poorly designed ASR system will have a negative impact on customer satisfaction and call center performance. The IVR should be viewed not so much as a call center automation opportunity, rather as another customer communications channel—a customer touch point—and managed as such.

The most popular IVR self-service application is not surprising—account inquiry, something that is common to all industries.

As shown in Figure 3.13, participants offer three basic IVR and Automatic Call Distributor (ACD) configurations to handle incoming call flow.

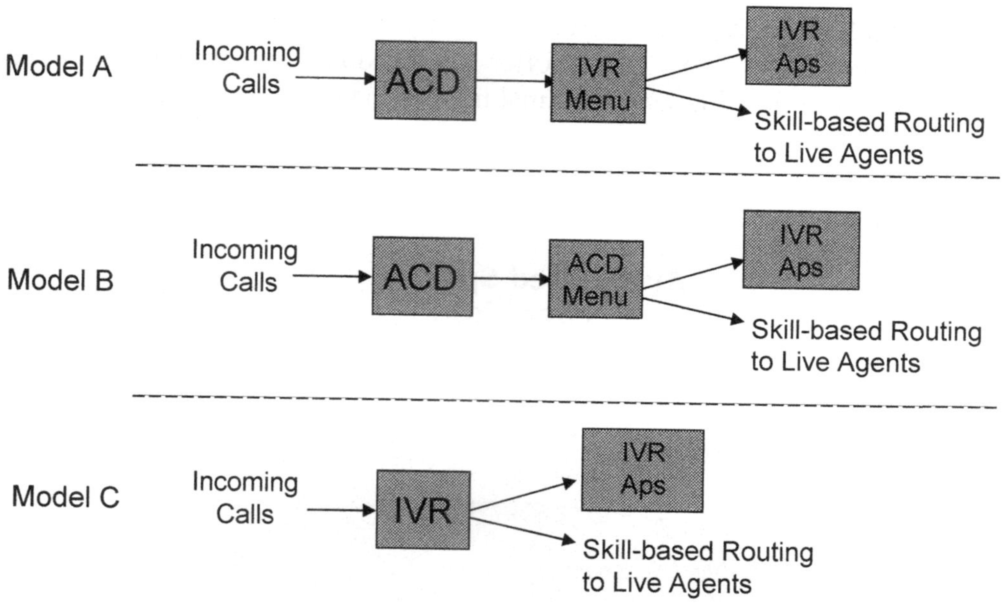

Figure 3.13 IVR configurations

The most popular configuration, demonstrated by 46 percent of survey participants, is Model A.

Half of the IVR deployments "selectively force" callers through the IVR for some applications—account balances, payment extensions, or outage reporting. This means that callers are not given a choice between an agent and self-service when selecting these options—callers are automatically placed in the IVR application and have to complete the transaction in the IVR or "opt-out" to speak with a representative.

Forcing callers through IVR applications increases system utilization and success, especially for callers who are unfamiliar with system options and functions. Selective forcing can encourage higher IVR usage for the more routine tasks while freeing up representatives to handle the more complex service issues. Selective forcing can also be dynamically deployed. For example, companies can selectively force callers into IVR applications during peak calling periods or off-hours, but allow callers to choose during slower periods. A smaller percentage of IVR deployments (10 percent), however, use the IVR primarily as a back-end, over-flow service—during peak calling periods.

Several participants have the capability to send a percentage of calls to an outside vendor's IVR, transparent to the customer. This option is usually used during heavy call volume to supplement existing in-house resources. Other participants have deployed additional back-up IVR capability in-house to handle these situations.

Forty percent of participants have deployed speech recognition software within their IVR and another 32 percent have near-term plans for speech recognition (Figure 3.14). Sixty-eight percent have deployed Computer-Telephony-Integration (CTI) in conjunction with their IVR (Figure 3.15). CTI refers to the integration of computers with phone systems to better handle calls. For example, while a call is in the queue, the computer retrieves customer account information and provides it to the CSR's screen as soon as the customer's call is answered. Both automatic speech recognition and CTI offer tremendous opportunities to streamline conversations and reduce operating costs, by making call handling more efficient and in some cases eliminating human involvement in a call.

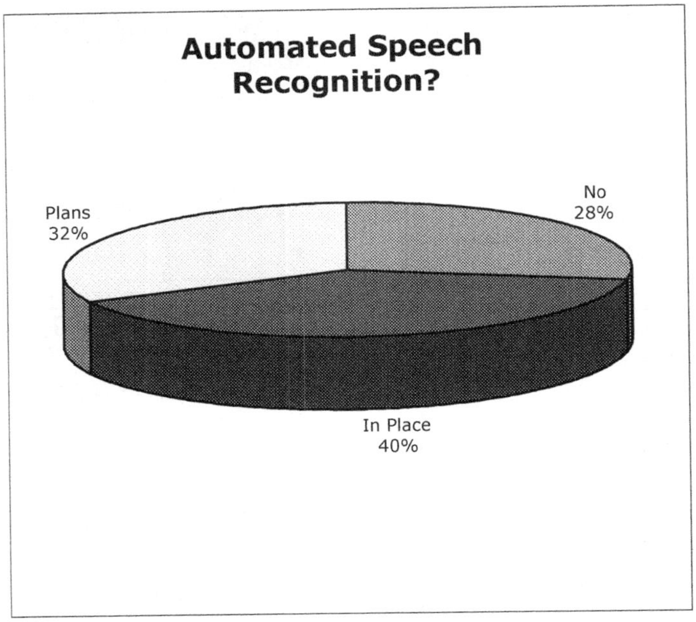

Figure 3.14 Participants using automated speech recognition

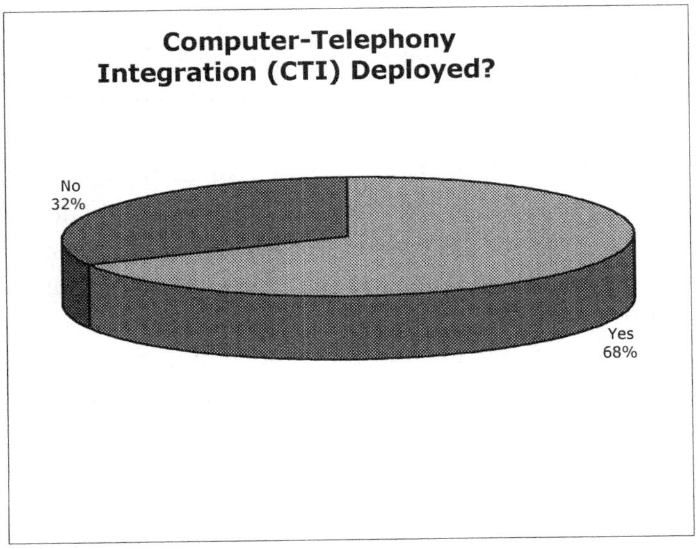

Figure 3.15 Participants using Computer-Telephony-Integration (CTI)

As shown in Figures 3.16 and 3.17, only 36 percent of participants offer wait time estimates to callers in the hold queue, and only 27 percent of participants promote IVR self-service options to callers in the hold queue. While these features don't impact the IVR directly, companies can increase customer satisfaction through the announcement of accurate wait times to callers in queue or by offering an alternative to waiting in queue, such as self-service through the IVR, or even a call-back scheduled by the IVR.

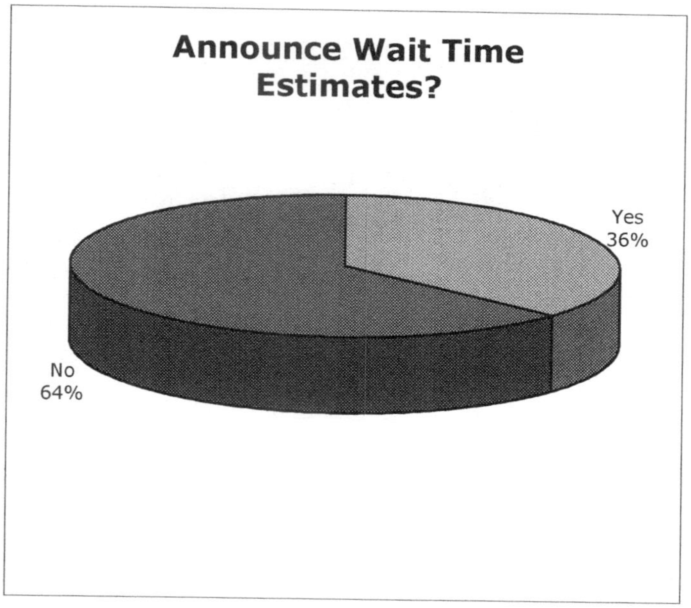

Figure 3.16 Participants that announce wait time estimates

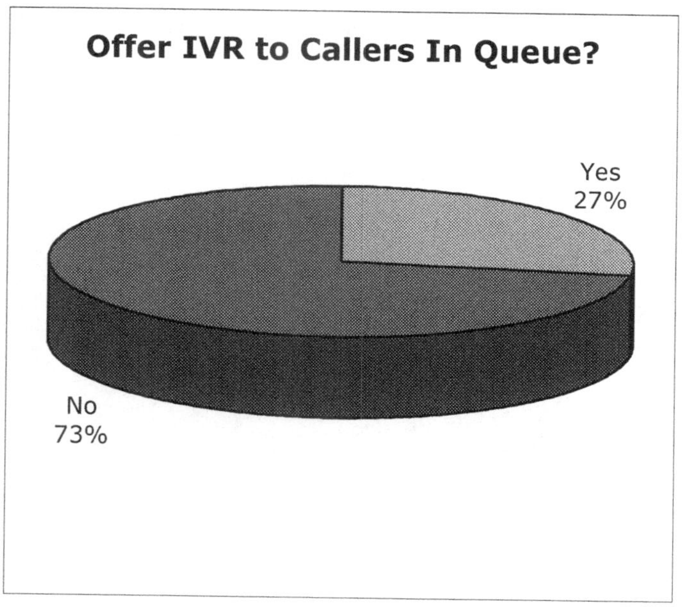

Figure 3.17 Participants that offer IVR to callers in a queue

One of the weakest links appears to be measuring IVR customer satisfaction. Only half of participants measure customer satisfaction with the IVR, in some form or fashion (Figure 3.18). Less than ten percent mentioned customer satisfaction as a measure of IVR success.

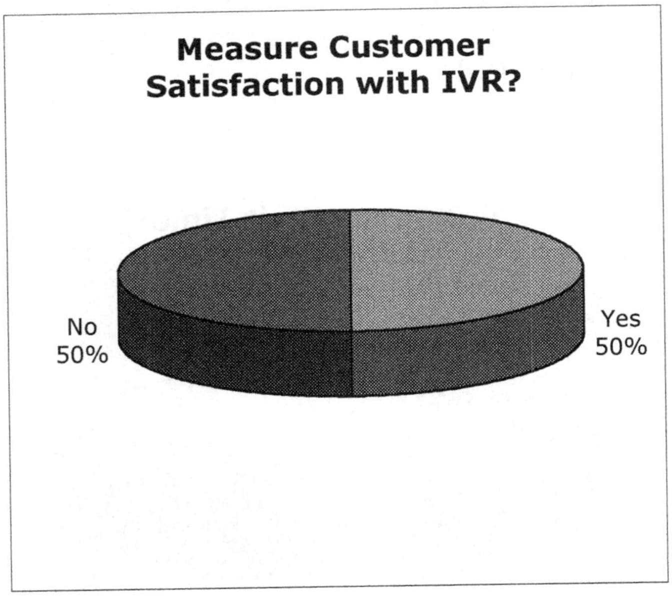

Figure 3.18 Participants that measure customer satisfaction with their IVR

Otherwise, companies reported the use of usability-labs, customer focus groups, call monitoring, internal pilots, and agents as a source of customer input and feedback for IVR design changes. One company reported comparing its IVR scripting and menus to that of its highly rated peers (based on customer satisfaction ratings).

Only 30 percent of participants promote IVR usage with customers through any means (Figure 3.19). The most popular form of promotion are bill stuffers or bill/envelope messages.

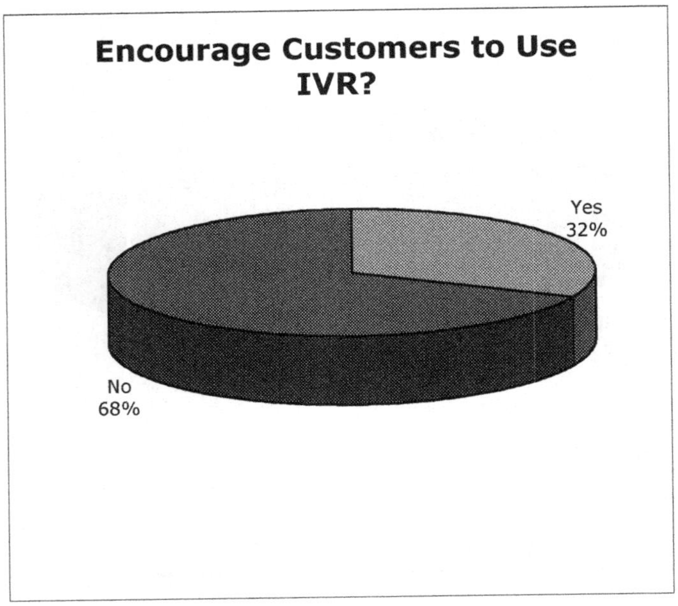

Figure 3.19 Participants that encourage customers to use their IVR

METRICS AND BEST PRACTICES

Key efficiency and effectiveness metrics were calculated from data supplied by 29 utilities, including 11 water utilities. While other types of organizations could have been included in the metrics analysis, we feel that utilities are more comparable for this purpose (but some practices mentioned in this chapter are from non-utilities, especially if they have high customer satisfaction). The figures below contrast the best performers (those utilities that had both above average efficiency and effectiveness) with the rest.

The key metrics chosen to provide a balance between service and efficiency are:

- Cost per call – Operational & Maintenance costs only – direct labor, contractor costs, overtime, and non-labor O&M; no capital costs or overheads
- Average Speed of Answer (Actual) in seconds
- Calls handled per FTE involved per day
- Service Level Conformance (% days service level goal met)
- Abandonment Rate (% abandoned calls or total offered)
- First Call Resolution Rate (% calls resolved on first contact of total handled)
- Agent Availability (% time agents on phone or available to take calls)

To maintain a balanced view of performance, utilities were designated as "best performers" if they exhibit above average performance in both cost and service. Once the "best performers" were identified, "best performer average" metrics were calculated, as reflected in Table 3.2 below.

Table 3.2 Call center benchmark performance.

All Utilities	Best Performer Average	Rest of Industry Average	Water Industry
Cost per Call	$2.63	$6.54	$3.70
Average Speed of Answer (Actual)	38	75	58
Calls Handled per Agent	71	49	61
Service Level Conformance	72.3%	67.5%	67.7%
% Abandoned	4.6%	9.2%	5.8%
% Resolved 1st Contact	81.2%	76.2%	94.5%
% Agent Availability	88.8%	76.0%	74.3%
Number Participating	5	29	11

CHARACTERISTICS OF BEST PERFORMERS

Companies that are achieving high customer satisfaction and high call center performance exhibit some common characteristics, as described below.

First Call Resolution

Best performers:

1. *Measure and Track FCR Performance at Multiple Levels (i.e. by agent, work group and entire call center).* FCR is a key determinant of customer satisfaction. Measuring FCR performance by agent provides actionable information that is useful in determining coaching and developmental needs. It also makes agents more aware of the importance of FCR and holds them accountable for their performance.

2. *Promote and Emphasize First Call Resolution and Utilize it to Drive Improvement at All Levels.* These organizations routinely discuss FCR performance and the impact of repeat calls and rework. They build FCR into new hire training, refresher training, and remedial training, review FCR performance during coaching and other one-on-one sessions, and measure FCR performance by call type, by agent, by team, by work group to gather as much information about customer service delivery as possible. They use the results to drive improvement by identifying call types and supporting work processes that need improvement.

3. *Give Agents the Proper Training, Authority, and Tools to Encourage Higher First Call Resolution.* They help agents to solve the customer's problem the first time.

4. *Incorporate FCR Measures into Employee Compensation and Incentive Plans.* As with any measure, there needs to be a balance with other measures to make sure all around good performance is delivered and service isn't compromised in the pursuit of a single measure, and that agents know when to escalate a call or assign it to further investigation.

Call Quality Monitoring

Best performers:

1. *Determine customer value and expectations to truly measure the customer experience.* Measuring the customer experience is the primary purpose of a call quality monitoring process. Best performers use customer research—interviews, focus groups, expectations surveys, or even calibration sessions to understand how customers' want to be served. Expectations vary from segment to segment and change over time, so calibrating call quality against customer expectations is an ongoing effort.

2. *Monitor, evaluate, and discuss results.* Some companies set up an in-house quality assurance group. There are also outsourcing options available from companies that specialize in call quality monitoring. Contracted monitoring services can be provided on-site or remotely, based on individual specifications and standards.

3. *Monitor the call monitors routinely to build consistency, confidence, and monitoring skills.* One sure way to improve monitoring skills and bolster confidence is to monitor the

individuals responsible for call monitoring. Best performers listen to the calls that the monitors evaluate, compare evaluations, and coach the monitors.

4. *Train coaches.* Time and experience as a representative does not necessarily develop coaching skills—employees must be trained. Customer service leaders take the time to develop and implement a proper training program to ensure that supervisors, trainers, managers, and QA personnel are equipped to provide constructive feedback and developmental guidance.

5. *Develop and evaluate coaching performance,* including performance measures to track team progress and performance and hold coaches accountable for their team's improvement.

6. *Communicate and educate agents on the monitoring processes,* especially how calibration sessions develop consistency and how monitors are held accountable for the evaluations they render. Leaders also track and promote the successes achieved as a result of monitoring efforts.

7. *Let employees develop a better understanding and respect for call monitoring* by experiencing the monitoring process first-hand. These sessions also provide opportunities to assess monitoring skills in representatives for career development. Companies using this practice report that representatives are usually harder on themselves than the supervisor reviewing their performance.

8. *Use a variety of call monitoring techniques.* Many rely on recorded sessions in combination with live monitoring sessions, whether side-by-side or remote. QA groups tend to rely on remote monitoring using recorded voice and data while supervisors conduct more traditional live monitoring sessions—using silent and side-by-side techniques.

9. *Report that regular QA group calibration, and calibration with supervisors, are key drivers of QA organization credibility.* Other drivers mentioned include effective communication, QA participation in team/center meetings and training, obtaining input and feedback from representatives, and gaining management support for the QA program. The most frequently cited benefits of a QA program include the ability to identify individual employee developmental needs, improved consistency and call handling, and increased customer satisfaction.

Integrated Voice Response Deployment

Best performers:

1. *Reflect customers' values and expectations in their IVR systems.* Customer research through phone and mail surveys or focus groups are being used by best performers confirm fundamental needs and expectations. Further testing of IVR prototypes and subsequent enhancements helps to ensure that customer needs are met.

2. *Conduct extensive customer research in conjunction with any IVR design changes or additions.* Best performers use customer focus groups and other research methods to test

design prototypes, validate scripting and prompts, and confirm menu options, every time a change or enhancement is considered.

3. *Find that Speech Recognition can dramatically improve IVR utilization, however, its implementation requires a different approach than touch-tone.* Customers really like speech-enabled applications if it's done correctly. Users' experience suggests that the addition of speech to IVR applications has increased IVR utilization significantly. The benefits of voice recognition-driven applications are enticing—higher usage, shorter call lengths, natural conversation interaction, and increased customer satisfaction.

4. *Train Agents to Promote IVR Services.* By far one of the most successful, and probably least expensive, educational efforts a company can undertake is to encourage call center agents to routinely suggest that callers access and complete transactions within the IVR. When appropriate, agents inform callers that the transaction just completed with the agent could have been completed, without a wait, through the IVR.

5. *Selectively Force Callers through the IVR.* These companies average the highest IVR Success Rates—more than 40 percent (percent of IVR completed calls versus total handled calls). Conversely, companies that present IVR services as "optional" averaged an IVR Success Rate of only 13 percent.

6. *Measure and Track IVR Performance*, so that it's easy to track call flow, errors, voluntary and involuntary opt-outs and opt-out points, hang-ups, and success rates. Service leaders track IVR performance from the customer's perspective—how it impacted what he or she was trying to accomplish. Several vendors offer "stress test" software to assist here.

7. *Take advantage of a service or tool to actively test and monitor IVR applications.* These services are being used by service leaders to point out application, network, and system inefficiencies and failures that the organization may not recognize or hear about from customers.

8. *Monitor IVR Call Quality.* The technology is handling some of the same calls as agents, so the need is to monitor the call quality of IVR transactions similar to agent-handled calls. Call quality experts record and review IVR transactions to better understand the level of service that the IVR is offering customers. This is especially critical for speech-enabled systems.

9. *Target Research and Promotional Campaigns at those Callers who Voluntarily Opt-Out of the IVR or Frequently Talk with Agents.* This research is being used by service leaders to yield insight that will better the design of the system or increase its functionality. Agents can also be trained to point out the instances when it makes sense to use the IVR while the targeted campaign can reinforce the benefits.

CASE STUDIES

Lexus

It's not your typical call center. Call Lexus customer service for assistance and a single Customer Service Representative (CSR) stays with you until the case is resolved. Lexus' approach, known as "Case Management" provides the exceptional service its customers have come to expect. Lexus' call center maintains training vehicles—at least one of each Lexus model car. This enables representatives to familiarize themselves with each car's features and research specific inquiries. A rotating schedule allows representatives to drive each of the models. Lexus expects each representative to be so familiar with each model's operation that customers will hear, "Yes, I know exactly what you're talking about."

Candidates undergo a telephone screening before Lexus invites them into the company for a series of interviews. Representatives from human resources and the call center meet with the candidate, and then within the call center, Lexus conducts two panel interviews, with panel interviewers consisting of two Lexus employees, usually a supervisor and the department trainer.

New hires participate in four weeks of classroom-based training where they learn their "soft skills"—policies and procedures, computer system, and technical training related to the vehicles to familiarize the representative with the automobiles. Lexus requires no automotive knowledge; however, knowing basics, such as common automotive parts, assists the representative in answering customer inquiries.

During training, each representative is assigned a mentor. As training progresses—from day one—representatives spend a portion of each day sitting with the mentor, listening to phone calls, observing the representative handle the process. This assists the new agents in putting the process together. Training also involves a great deal of role-playing. By the end of the initial four-week training period, new CSRs type all the information into the computer, but observe and listen as the mentor speaks with the customer. When the new reps graduate to the floor, they are teamed with another representative or their team supervisor as they handle the first few calls and feel comfortable handling inquiries on their own.

Lexus requires a commitment of 18 months within the Center. After that time, associates are free to look for positions elsewhere within the company. By encouraging movement within the organization, Lexus avoids representative burnout. The CSRs learn so much about the dealerships and the products that they become prime candidates for other positions within the company. Many associates advance to the field as district service managers or eventually district sales managers.

What Makes Lexus Uniquely Lexus? It's the people in the call center who make Lexus tops in customer satisfaction, and their ability to do anything for the customer. Lexus agents look for ways to say yes. The secret is in what agents do for the customer, not in the systems or technology—just listening to the customer and hearing what they need, from their perspective, and determining how to exceed expectations.

Pepco

Pepco is an investor-owned utility that provides electricity in Washington, DC and the Maryland suburbs, serving more than 700000 customers. For the past few years, J.D. Power and Associates has ranked Pepco among the top energy utilities in the eastern U.S. in terms of customer satisfaction. Pepco earned further recognition in 2002 when its Customer Care organization was honored by Call Center Magazine as one of the Best Call Centers in the nation and its customer service manager, Charles Dickerson, was named Best Manager of 2002.

Pepco relies on an Avaya switch and Avaya's Best Service Routing (BSR) to steer callers to the right agent. Pepco used to train every agent to be a super-agent. This new technology offers the flexibility to ease new agents into the job much sooner. Pepco has structured BSR to select the next available agent based on several factors, including the available pool of agents, the type of call, and agent talk time. These routing rules make sure that agents who just finished a lengthy conversation are not immediately facing another caller.

Pepco has also implemented Blue Pumpkin's workforce scheduling and management system. The combination of BSR and workforce management has helped Pepco meet its service level goals more efficiently.

Pepco's call center organization received a makeover when the company moved its headquarters from Pennsylvania Avenue, Washington D.C to Forestville, Maryland. When Pepco's main call center relocated to the Forestville facility, management took advantage of the move to create a comfortable and inviting work environment for its employees. Employees actively participated in the decision to relocate the call center. Management surveyed employees to determine the amenities deemed important and to identify any employee concerns regarding the move. This involvement achieved buy-in and made the move to the newly renovated center a tremendous success.

In addition to the new furniture and ergonomic workstations, Pepco management added a "stress relief room" and lunchroom to the call center. The stress relief room is equipped with exercise bikes, treadmills, and other workout machines that help employees get the kinks out before or after a shift, or even during a break. According to management, the stress relief room has had a big impact on agent morale.

Employees are the key to making Pepco successful. The call center engages its employees on a continuous basis, regularly conducts team meetings, involves employees in new projects and initiatives, and keeps them abreast of its business plans and strategies. In addition, call center employees meet frequently with the vice president of Customer Care in a "brown bag" lunch forum—giving employees the opportunity to discuss issues with senior management.

Pepco management has also assembled a team of CSRs, supervisors, and managers to address ongoing issues and challenges—the Cooperative Improvement Committee (CIC). Committee members represent all levels of employees within the center and work together to resolve issues and create solutions that will work for everyone.

(continued)

Pepco (continued)

Most recently, Pepco introduced a "virtual hold" service that offers callers the choice of holding for the first available agent or requesting a return call, without losing their place in the call queue. Pepco is working to improve the customer experience and at the same time improve its call center performance through reduced hold times, fewer abandoned calls and improved employee performance. When a customer calls Pepco and the wait time is expected to exceed two minutes, the system announces the estimated wait time and offers the caller an option to receive a return call, instead of waiting on hold. When the caller selects a return call, the system prompts for a callback telephone number and asks to record the caller's name. The system then provides a callback time and disconnects. The system keeps the caller's place in the queue and calls the customer back shortly before its time to transfer the caller to the first available representative. Customers are thrilled with the technology and the ability to reach a representative, even during busy calling periods.

Pepco management has now turned its attention towards first call resolution. Management introduced its first-call resolution initiative with special events and activities in the Call Center and Business Offices to educate, engage, and involve employees in achieving the goal. In addition, the call monitoring evaluation form has been modified to accommodate the goal, performance standards have been adjusted, and several recognition incentives have been developed to reward positive performance.

Pepco established a subset of CSRs to serve as a liaison between Customer Care and the Distribution Dispatch Center. Recognizing the importance of effective communications, especially during large outages and storms, Pepco cross-trained 15 to 20 CSRs to keep the Center updated on storm situations and work order progress, as well as to handle special safety-related transactions. The increased skills and responsibilities offer agents developmental opportunities and a break from time on the phones.

Pepco recently streamlined the menus and prompts on its Integrated Voice Response (IVR) technology, shaving a few seconds off total answer time and saving the company thousands in toll charges. Customer satisfaction with the IVR rose by 3.5 percent after these changes. Management plans to further enhance the IVR with speech recognition. Adding speech recognition will make it easier to automate additional transactions, such as processing phone requests to pay by check, credit card, or debit card.

First Direct

First Direct is a traditional and on-line retail bank. When First Direct launched on October 1, 1989, branchless telephone banking was untested and to the majority of people, a completely new idea. Today, the impact of First Direct extends throughout the entire financial services industry. Offering direct financial services is no longer unusual and far from the choice of the few. It is the preferred method of banking for a growing number of the UK's 38 million banking customers. First Direct offers personal banking only—no business banking.

(continued)

First Direct (continued)

Representatives answer more than 300,000 calls each week—First Direct does not use Integrated Voice Response technology. On average, representatives answer calls within 10 seconds. Representatives handle customer inquiries regarding bank account balances, fund transfers, checks and debits cleared, and other banking matters. CSRs, when needed, will refer calls to other departments for resolution; however, customers may be transferred only once.

First Direct actively promotes a positive working environment, believing that high levels of satisfaction amongst its employees translates into a more enjoyable banking experience for the customer. A nursery operates at each of the three sites as well as a subsidized restaurant. A sports and social club organizes a wide range of activities, including sports teams, nights out, day trips and discounts at local shops and services.

First Direct actively encourages career development. Employees can pursue opportunities to transfer across the differing areas of the business, from IT to personnel.

Recognizing the importance of people making the Company what it is today, First Direct gives recruitment and training top priority. First Direct's approach to recruiting customer-facing people centers on employing individuals with excellent communication skills and who enjoy serving customers. First Direct has determined that it is easier to teach the necessary skills, than to teach employees how to enjoy working with customers.

All new recruits go through an intensive five-week training course, which provides an in-depth examination of products, communication techniques, and most importantly, how to listen to the customer. After the initial training, employees undergo 18 months of ongoing coaching and accreditation.

Representatives are encouraged to "go the extra mile" to satisfy customers. As part of its training program, First Direct arms employees with the appropriate skills and knowledge to encourage on-the-spot decisions. Agent teams can award a "green sheet" to another team for helping out in a difficult situation or unusual customer scenario.

In terms of customer satisfaction measurement, First Direct conducts regular surveys over the telephone, Internet, and face-to-face. The purpose of these surveys is to discover how customers feel about all aspects of the bank, from its services and marketing, to the types of products that people would like to see offered. Extremely service-oriented, First Direct recognizes that customer satisfaction drives growth through referrals.

First Direct exists for the customer's convenience, not its own. It doesn't designate any hours as premium—no hours are more important than others. For instance, employees who work the night shift get the same pay as those working the day shift. For First Direct, this helps maintain its cultural integrity—to prove that each hour is important and that it is the customer who matters.

First Direct fundamentally believes that each customer is an individual and must be treated as such. The same applies to its employees. First Direct is single minded in delivering first class service, 24 hours a day, seven days a week, 365 days a year. First Direct is determined to ensure that service remains at the heart of the online experience and that its people are committed to setting the standard for e-service in the future.

Louisville Water Company

Louisville Water Company (LWC) is a city-owned drinking water utility, serving 800,000 people in Greater Louisville and surrounding counties. LWC has 16 employees in the call center, which handles an average of 1,700 calls per day. The CSRs process phone calls relating to move orders, service requests, water quality, billing inquiries, payment extensions, adjustments, credit/debit card payments and general information. A Symposium ACD system collects data on the numbers of calls received, types of calls, length of calls, real-time statistics on CSR availability, queue, and abandonment rate. Calls are monitored on a random basis for quality assurance.

CSRs are selected based on various screening processes including behavioral-based questions.

CSRs are cross-trained with the cashiering and remittance processing staff. This helps the company to manage CSR staffing levels tightly and allocate employees flexibly between these functions according to daily and even hourly work volumes.

The abandonment rate for LWC remains low. This is reflected in the consistently high customer satisfaction ratings in its customer satisfaction survey (84% of customers are mostly or very satisfied with the company in recent surveys.)

CSRs are trained to move quickly through calls and to resolve most issues on the first call. There is a quick reference guide available to help them find the right staff member based on the customer's issue, if a referral is needed.

CSRs have the same authority levels as senior management within the corporate rules, so they do not often have to refer issues up the line, e.g. by dollar value.

Managers make sure distractions in the call center are minimized to keep productivity high, and minimize meetings involving CSRs.

Rewards are given monthly to CSRs who meet their goals, in the form of extra time off. CSRs have said they appreciate this above other rewards used in the past e.g. gift vouchers. There is no limit to how many CSRs can win the extra time off (in the past, the reward system was based on the highest performers, which demotivated other staff who had met their goals but were not in the "best" category).

Career progression is available within the call center based on skills. There are 3 levels of promotion possible within this function, which encourages CSRs to stay in the role rather than leaving the call center to get a promotion.

CHAPTER 4
METER READING

TYPICAL PRACTICES

This chapter focuses on meter reading metrics and practices, with an emphasis on key metrics and practices used by best performing organizations, and practical recommendations to water utilities. The utilities that provided data included in this study had the following service territory makeup (Figure 4.1).

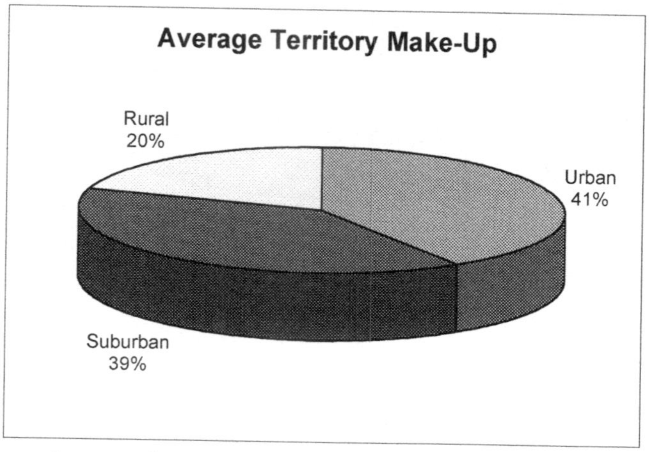

Figure 4.1 Average service territory make-up

Utilities are still mostly reading meters manually and are likely to be reading them manually for some time. While the use of automated meter reading technologies (AMR) is increasing, over 90% of water and other utilities are still reading the majority of their meters manually. Therefore, utilities are focusing on reading meters more efficiently and effectively.

At the same time, as a cost reduction and service improvement initiative, many utilities are strategically deploying AMR, especially in high-read-cost situations. The trend to using AMR will likely continue as its cost becomes more attractive for the bulk of residential meters. Recent technological innovations, such as Power Line Carrier (PLC) AMR, which can allow electric, gas and water meters to be read over electric power lines, have expanded the benefits of AMR, and are convincing more utilities to expand their AMR programs. Such benefits include the ability to answer most meter-related inquiries while the customer is on the phone.

Most utilities are focusing on one or more of the following approaches to meter reading improvement:

- Reducing costs of manual reads through contract negotiations, rerouting, more sophisticated hand-held equipment and meters, and lowering overhead. Some have reduced costs to a point that makes it difficult to justify AMR for residential accounts.

- Contract meter reading to reduce overhead, tackle seasonal peaks, or as a strategy to transition to automated meter reading.

- Automated meter reading – if not large-scale implementation, often to address high read cost meters, unsafe meter locations, and high-turnover premises.

Most utilities test new meter readers as part of the new hire training process. Close to 70 percent of utilities report the use of testing during the new hire training process (Figure 4.2). Testing is an effective technique to assess training comprehension and to determine readiness.

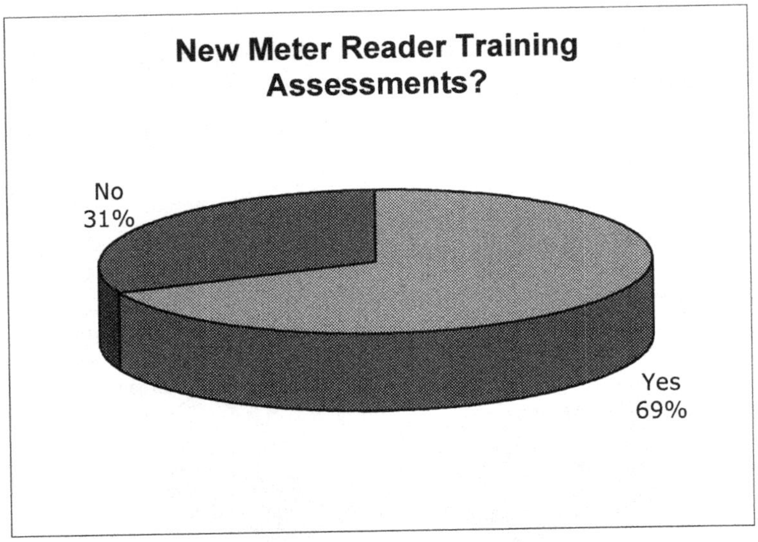

Figure 4.2 New meter reader training assessments

Computer-based training does not play a dominant role in meter reader training. Less than one-third of respondents reported the use of computer-based training in the new hire meter reading training process (Figure 4.3). However, more than 75% of respondents that offer refresher training incorporate computer-based training. Computer-based training offers a self-paced training program to help meter readers learn at their own pace, comprehend, and recall.

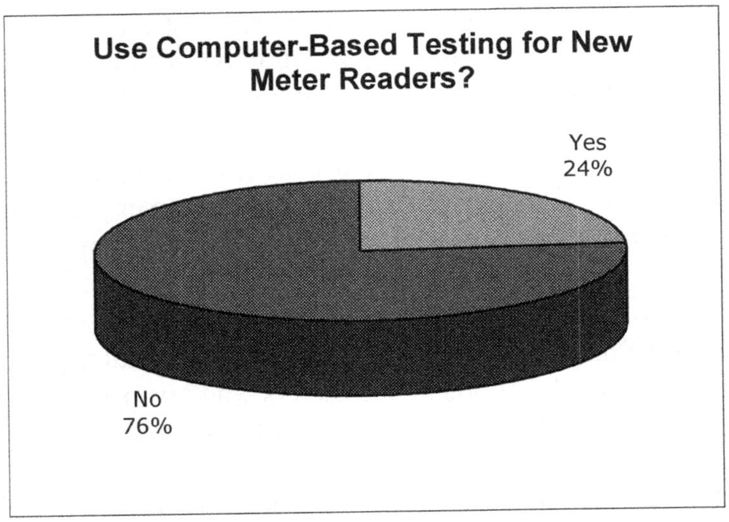

Figure 4.3 Participants using computer-based testing for new meter readers

The majority of participating utilities do not let meter readers go home after completing the day's assignment (Figure 4.4).

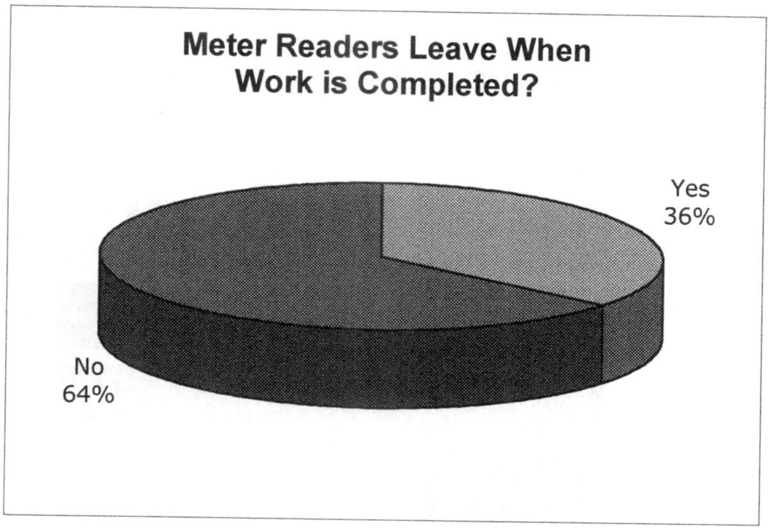

Figure 4.4 Participants allowing meter readers to leave when work is completed

Only a few companies read meters for other utilities, mostly small municipalities (Figure 4.5).

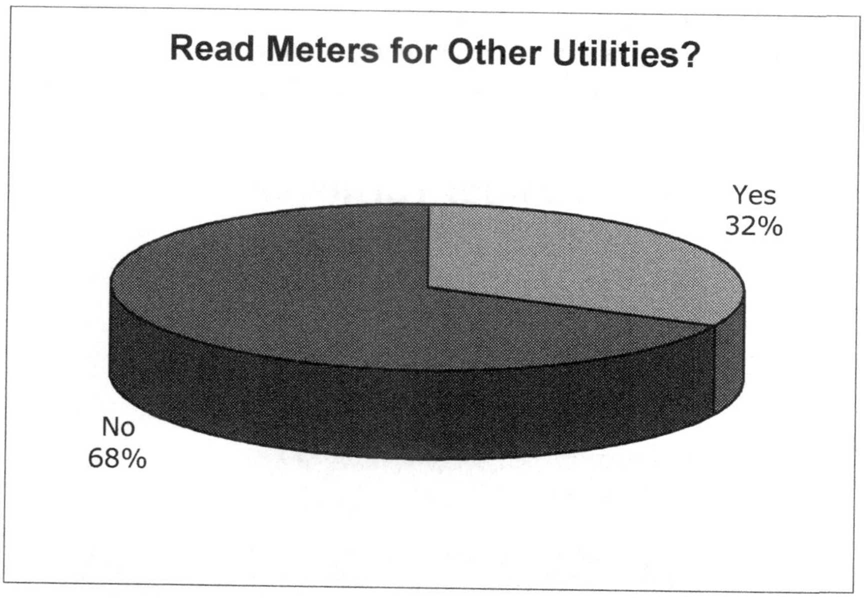

Figure 4.5 Participants that read meters for other utilities

The majority of participants reported an average length of service for meter readers of 8 years (Figure 4.6). Many participants indicated that meter reading was an entry-level position, and as such expected high internal turnover.

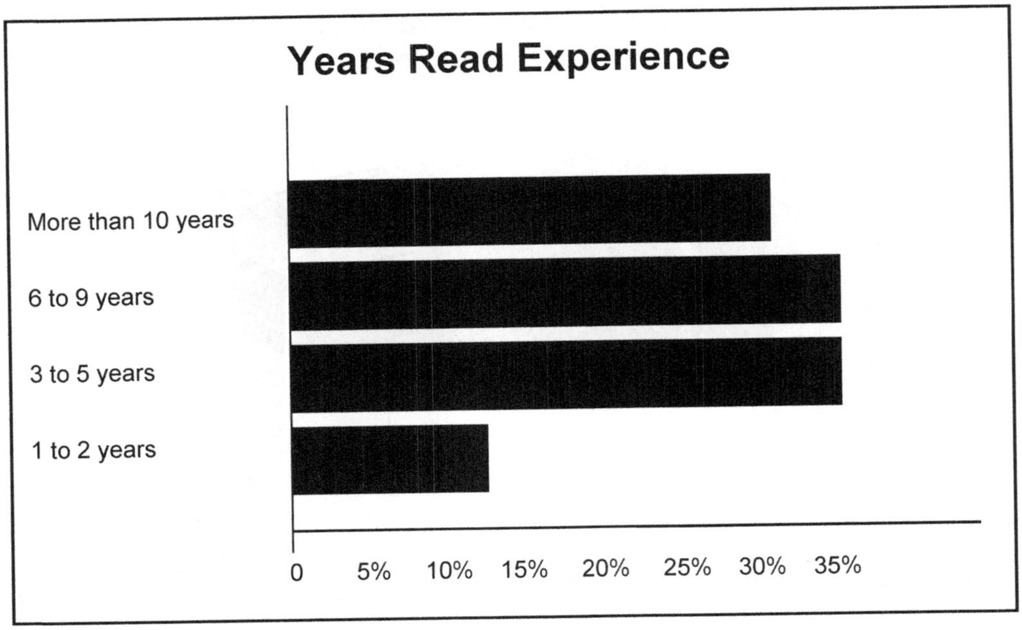

Figure 4.6 Years read experience

As a group, annual turnover averaged 14 percent. The highest turnover is seen in employees with less than 2 years read experience, and conversely, the lowest turnover is experienced in employees with more than 10 years read experience (Figure 4.7).

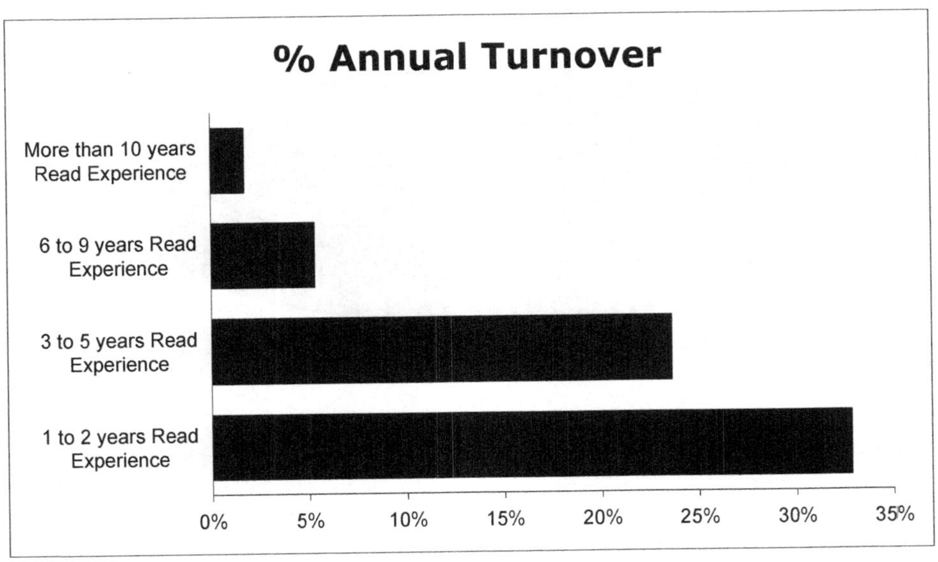

Figure 4.7 Percent annual turnover

Of the companies reporting average length of service of 10 to 20 years, several were in the process of fully implementing AMR and the rest of them use a higher paid position, field service representative, to read meters and work field service orders.

More than half of participating utilities use Route Management Software (Figure 4.8). Typical reported savings range between 10% and 20% with the use of this software.

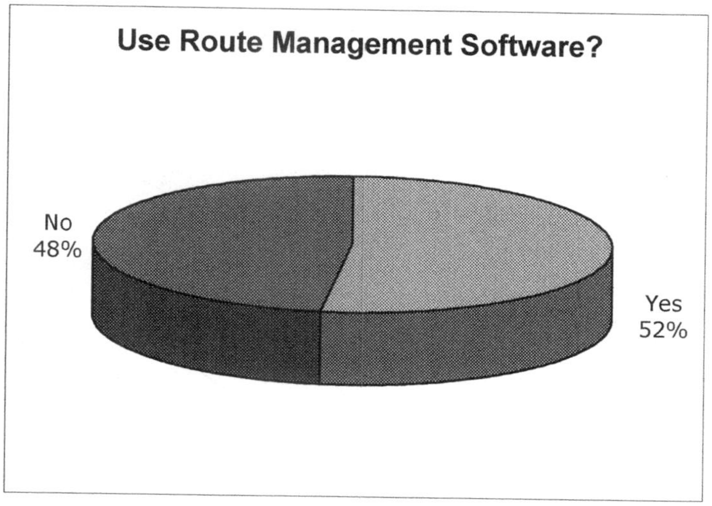

Figure 4.8 Participants that use route management software

Use of automatic meter reading among respondents is shown in Figure 4.9 below. Most utilities have no AMR meters deployed presently. As a group, participating utilities averaged 16% AMR meters. Those with AMR meters tend to have less than 20 percent AMR meters deployed. Electric utilities have the highest concentration of AMR meters.

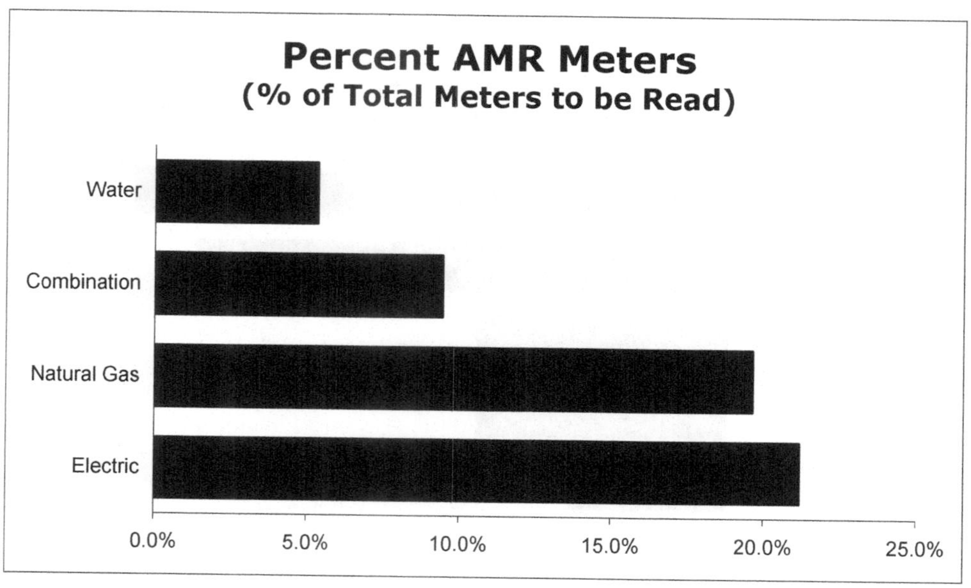

Figure 4.9 Percent AMR meters

Radio technology is the most common AMR Technology used today (Figure 4.10). However, power line carrier technology (PLC) is often favored in rural and semi-rural areas.

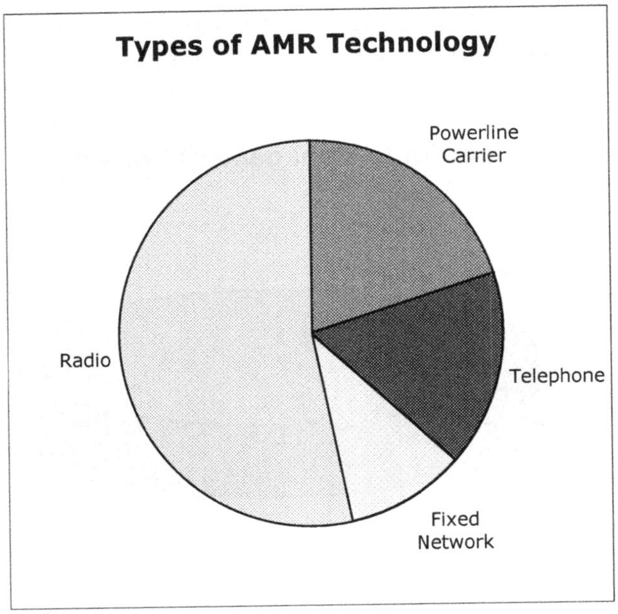

Figure 4.10 Types of AMR technology

Participants were asked to identify the measures used to evaluate meter reader performance. As seen in Figure 4.11, route completion and read accuracy are the top performance measures used. Other measures used include customer relations, teamwork, complaints, amount of time worked daily, conduct, job knowledge, resource management, communication, and relationship with supervisor. Surprisingly, nearly one-third reported no performance measurement system in use.

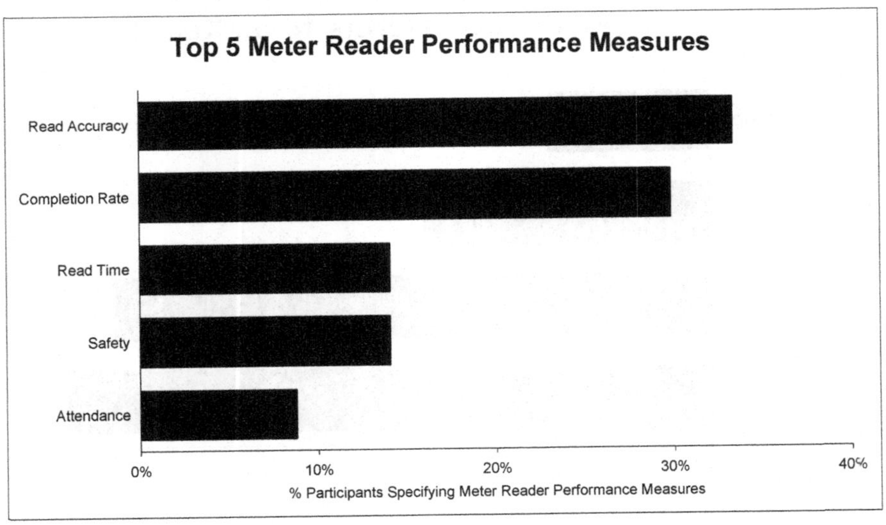

Figure 4.11 Top 5 meter reader performance measures

AMR is the most popular technique to address chronically inaccessible meters (Figure 4.12). Other techniques include supplying customers with read schedules, calling customers the night before the read, reversing routes every other month, Saturday reads and special skip routes.

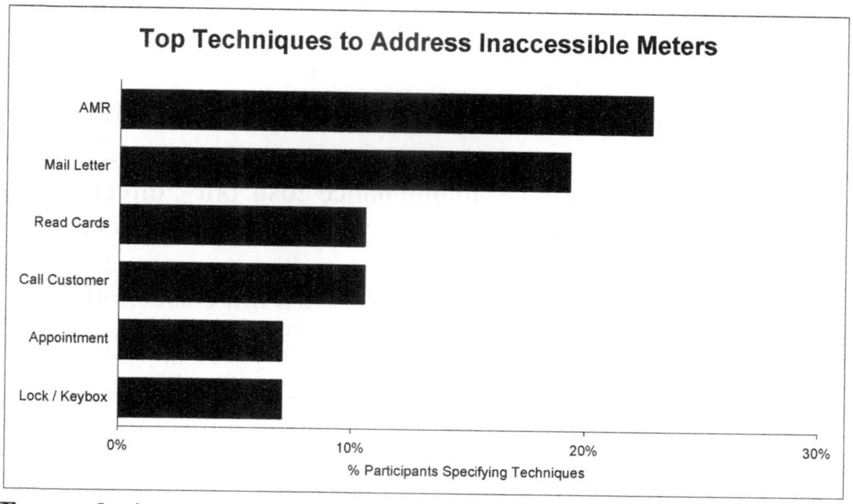

Figure 4.12 Top techniques to address inaccessible meters

Pepper spray was the most used dog control technique (Figure 4.13). Other dog control techniques include focused training, dog sticks, umbrellas, and dog biscuits. The effective use of dog control techniques can help reduce the percentage of re-reads and lower operating costs.

Several vendors offer telescoping devices aimed at disarming or distracting a dog or keeping a dog at a distance. Several companies reported using videos, conducting yearly refresher training, as well as including dog handling in initial meter reader training. Other utilities reported working with local humane societies and other associations to schedule dog control demonstrations with live animals.

Many utilities will code the meter as a "bad dog" location to warn the meter reader and to initiate other approaches such as customer assistance and special access, read with a monocular, remote (from the meter) dials, or automated meter reading. "Bad dog" meter locations are also often avoided and the account is estimated until the customer offers resolution.

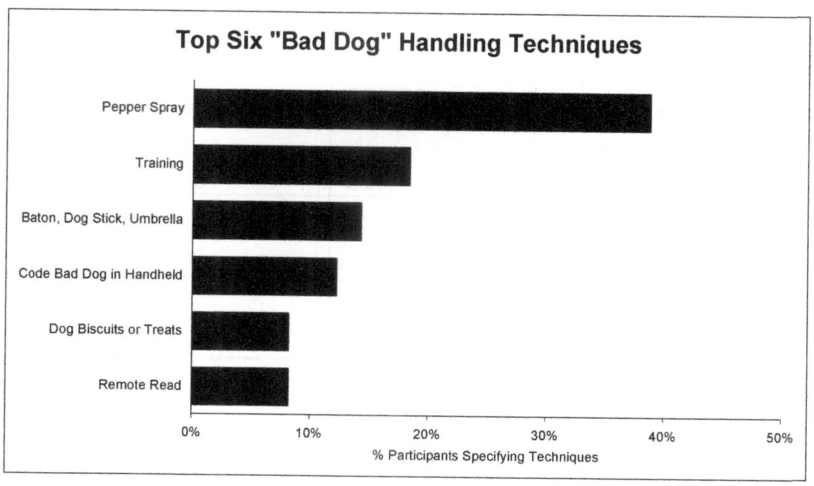

Figure 4.13 Top six "bad dog"-handling handling techniques

METRICS AND BEST PRACTICES

Key efficiency and effectiveness metrics were calculated from data supplied by 56 utilities, including 15 water utilities. Table 4.1 below contrasts the best performers (those utilities that had both above average efficiency and effectiveness) with the utility and water utility industry averages.

The key metrics chosen to provide a balance between service and efficiency are:

- Cost per meter read (operational & maintenance costs only: direct labor, contractor costs, overtime, and non-labor O&M; no capital costs or overheads)
- % Accurate readings (i.e. no misread that leads to a bill correction, for a representative month)
- Meters read per FTE involved per month
- % Meters read (of the total that should be read, for a representative month)

To maintain a balanced view of performance, utilities were designated as "best performers" if they exhibit above average performance in both cost and service. Once the "best performers" were identified, "best performer average" metrics were calculated, as reflected in Table 4.1 below. "Participating utilities" indicates the number of utilities that are used in the calculations for these metrics. In the sample, there were 7 "best performers" identified out of 56 utilities participating, of which 15 were water/waste water utilities.

Table 4.1 Meter reading benchmark performance

All Utilities	Best Performer Average	Utility Average	Water Industry Average
Cost Per Read	$0.38	$0.53	$0.67
Meter Reading Accuracy	99.96%	99.73%	99.16%
Meters Read per FTE	11,320	9,000	8,999
% Meters Read	99.7%	95.1%	95.0%
Participating Utilities	7	56	15

CHARACTERISTICS OF BEST PERFORMERS

Best performing companies are strategically implementing AMR, continually optimizing routes, effectively training, and encouraging high performance. Employee performance in best performing companies is measured both at the level of the individual and for the meter reading group. Employees have a clear idea of job expectations and performance, reinforced through performance metrics, rewards, and incentives. More specifically, best performing companies are:

1. *Using Automated Meter Reading (AMR) strategically to address access-problems, high-read-cost meters, unsafe meters, and high turnover premises.* Strategic deployment of AMR technology is an effective way to reduce cost, improve safety, and increase customer satisfaction. Best performing companies are utilizing AMR to tackle the problem-meters and high-read-cost meters first. In preparation for deregulation, utilities are also implementing AMR for large businesses and commercial customers. While a company-wide implementation may not be feasible, a strategic deployment to address problem meters can be very effective.

2. *Rerouting continuously* to *optimize routes, especially during the transition to AMR.* As long as there are routes to be read, whether manually or not, there will be room for optimization. Best performing companies report continual rerouting and route optimization to maximize productivity and reduce costs. A 10 to 20 percent gain in efficiency is achievable with a company-wide rerouting. Rerouting is also critical in areas of high growth and after an acquisition or merger. Companies with AMR implementations stress the importance of route consolidation and optimization throughout the automation.

3. *Implementing clear and concise measures of meter reader performance* – "Best performers" are deliberate in the measurement of employee, group, and departmental performance—cost, service, and productivity. Employees are given a clear idea of job expectations and performance and receive frequent performance feedback.

4. *Encouraging high performance through incentives and rewards* – Programs vary from bonus pay, special recognition, gift certificates, "bucks" redeemable at the company store, steak dinners, and other non-cash awards.

5. *Reducing Time-to-Standard through Effective Classroom and On-the-Job Training.* Best performing companies reported an average of 1 to 2 days of classroom training and 12 to 13 days of on the job training. Best performers' meter readers reach their company's standard 20 days sooner than average. These same companies reported the fewest errors and skipped meters, the highest productivity, in combination with the lowest unit cost. Companies reporting shorter than average training programs tended to have higher error rates, skip rates, and increased unit cost. Training is also recognized as a key factor for success in reducing error rates, detecting tampering and lost meters, improving safety, and improving customer service.

CASE STUDIES

PPL Electric Utilities

PPL Electric Utilities (PPL) is an investor owned utility serving about 1,000,000 people in Lehigh County and adjacent areas near Philadelphia. PPL has implemented an AMR system covering virtually every customer. It is one of the first large-scale implementations of Power Line Communication (PLC) technology. The system sends meter reads and has 2-way communication with meters via power lines.

PPL realized 3 types of benefit: tangible, intangible and future. The business case was justified on tangible benefits only, which were: 1) elimination of manual meter reading, i.e. labor, vehicles, etc.; 2) elimination of estimated reads which reduced the call center workload considerably; 3) ability for high bill complaints to be resolved by the CSR while the customer is on the phone, as the CSR can ping the meter and also examine daily readings, saving many special meter reads; 4) improved accuracy of large meter reads and elimination of manual checks; 5) improved outage diagnosis which can be carried out without sending a crew, i.e. PPL can establish from the call center whether it is the utility's or the customer's side of the meter that has the problem; 6) easy confirmation as to whether all customers are back on line after an outage, by testing all the meters in the affected area; 7) reduced short term meter maintenance costs, because all meters were recalibrated or replaced during the system implementation.

Intangible benefits include: 1) reduced theft of service, which is the biggest intangible benefit by far (national estimates are that 1-3% of electricity is being stolen, however by definition the amount at any one utility is unknown); 2) capacity to read meters hourly; 3) meters that can flag if they have been tampered with; 4) data mining of customers' energy use patterns; 5) improved metering accuracy; 6) better customer service and higher satisfaction, leading also to more favorable image with the PUC.

Future benefits are expected to include: 1) more optimized preventative maintenance, ability to make more precise decisions regarding periodic maintenance than just following fixed maintenance schedules; 2) possible support to more tailored tariffs if time of use charging can be increased; 3) more granular understanding of usage could improve wholesale energy purchasing decisions; and 4) more accurate financial reporting, less estimated usage.

The system is working well. PPL believes that the system would be especially applicable to medium-sized municipal utilities and co-ops in semi-rural areas. Change management was easier than other projects such as CIS, WMS and Y2K, because the AMR system made CSR's jobs easier, is a specific application without many of the integration problems of the other systems, and the AMR system was rolled out in phases.

Yarra Valley Water

Yarra Valley Water (YVW) is a water/wastewater utility serving about 1,000,000 people in Melbourne, Australia. YVW has outsourced its meter reading function for 10 years. In July 2000, Yarra Valley Water implemented a new "fixed price" contract based on an understanding of the total cost of meter reading, including IT support, back office follow up and other overheads. This made the contractor not only responsible for the collection of meter reading data but the quality of the data and any rework. The contract includes key performance indicators (KPIs) to enhance billing accuracy, and incentives for efficiency via bonus payments where KPIs are exceeded. In past "rate based" contracts there was little incentive to drive efficiency as they were paid to attempt a read and each subsequent visit was also paid for. This meant all the problems were sent back to the Company to resolve. Since the implementation of the "fixed price" contract, the "meter no read rate" has reduced from 4.9% to 0.56%, all 26 geographical areas (approx 22,000 meters each) have a no read rate of <1.0% and meters with >3 consecutive no reads has reduced from 1.5% to 0%. Reducing the meter no read rate and the subsequent need to issue estimated accounts is consistent with the principles of YVW's Customer Service Strategy. Estimated accounts cause unnecessary calls from customers, as they are dissatisfied with the accuracy of their bill, also increasing YVW's costs.

To assist meter readers, Yarra Valley Water is implementing a "virtual office" environment that no longer requires meter readers to report to an office every couple of days. Data is transferred directly to and from the hand-helds via modems installed in the meter readers' homes. Yarra Valley Water is also preparing to trial web-based meter reading using mobile phone technology.

Nashville Electric Service

Nashville Electric Service (NES) is a municipally owned electric utility serving the metropolitan Nashville-Davidson County, TN, area. The utility has about 296,000 customers. NES has successfully implemented a fixed network AMR system to address its highest turnover apartment complexes. NES's MicroNetwork, by Itron, combines radio and telephone communications technology to gather reads. Since its implementation in 1998, NES has saved significantly by virtually eliminating the once daily trips to the complexes to gather final reads for in/out orders. Now, reads are pulled by computer to process the on/off order in these locations. NES has also been deploying AMR technology to address hard to access meters. About 12 percent of NES meters have been automated through Itron's fixed network.

NES implemented a meter reader incentive system a few years ago to reward meter readers with good performance. Meter readers earn points through customer compliments, finding lost and tampered meters, and lose points through errors, accidents, and customer complaints. Meter Reader of the Month receives a cash bonus, special parking spot, and plaque. The Meter Reader of the Year earns a trip with his or her supervisor to the Annual Meter Reading Conference.

NES meter readers can go home when they have completed the assignment for the day, at 1pm Monday through Thursday and at 11am on Friday. Meter readers must read all meters on their routes, work skips and investigations, and maintain a low error rate.

Long Beach Water Department

Long Beach Water Department is a City-owned utility serving 460,000 people in Long Beach, California. Its meter reading is contracted out to Southern California Edison (SCE). Each billing day, the City transmits a file of meters to be read and changes to the meter locations to SCE. SCE's meter readers read water, gas and electricity meters at the same time. SCE charges $0.77 per read to the water utility.

The contractor reads a percentage of meters daily until the entire billing cycle is completed. The contractor completes 40 pickup reads daily. The contractor uses hand held devices to collect the reads.

The contractor submits a read file and an information update file back to the City.

The City reviews this file to determine what reads should be investigated prior to billing. The City uses meter readers to update records.

The contractor is allowed 2.5 errors per thousand reads prior to financial penalties being applied. The contractor currently performs significantly better than this level of errors.

The Department is currently renegotiating the per meter fee and also changing to bi-monthly reading (but monthly billing) to further reduce costs.

Orlando Utilities Commission

Orlando Utilities Commission (OUC) is a municipally owned public utility providing water and electric service to the City of Orlando and adjoining portions of Orange County. OUC is the second largest locally owned electric utility in Florida and the 16th largest in the nation.

OUC recently revised its Meter Reader pay structure, moving from salary to hourly pay, to provide more flexibility and opportunity for employees to earn overtime when it is required. This has been a morale booster for employees.

All meter reader job candidates must pass a hiring assessment that tests the ability to read a meter. Candidates must successfully read at least 95 out of 100 meters on the test. This ensures that OUC is hiring employees who can already read a meter or have the aptitude.

OUC implemented a new performance evaluation program to emphasize attendance, accessibility, and accuracy (3As). Employees are evaluated every six months. Meeting or beating standards qualifies employees for a step-increase in pay.

(continued)

Orlando Utilities Commission (continued)

OUC has implemented an incentive program within Meter Reading to encourage high productivity and accuracy. Meter readers earn OUC Bucks for high accuracy, attendance, and for finding lost meters, tampering, current diversion, stolen meters, and incorrect meter multipliers. Employees can spend the OUC Bucks at the company store or redeem them for gift certifications.

OUC optimized routes by completing a major rerouting of all meter routes in 2001 using TR*Star Automatic Route Management Software. OUC relies on a task assignment -- meter readers can go home after completion of their route and follow-up on any can't gain access situations (but they are still paid for 8 hours).

Every two years, OUC provides dog safety training with live animals -- through the local humane society or other organization. Employees learn the best ways to deal with dogs, watch videos, and discuss techniques. In addition, OUC offers dog-handling training through video.

OUC has deployed Itron's MicroNetworks to automate the reading of meters in apartment complexes. This has provided significant savings for the company. In addition, OUC is replacing all electric meters with Itron ERT meters to transition to AMR. Currently about 16% of meters are ERT meters. Additionally, OUC is piloting Elster's REX AMR technology to test its two-way communications capabilities in several locations within OUC's service territory.

CHAPTER 5
BILLING, PAYMENT, CREDIT & COLLECTION

TYPICAL PRACTICES

This chapter's focus is on billing, payment, and credit and collection metrics and practices, with an emphasis on key metrics and practices used by best performing organizations, and practical recommendations to water utilities.

Billing includes any task related to routine and non-routine billing by any channel, billing adjustments, rate application analysis, billing exceptions, customer research, and related records management/filing.

Remittance processing includes processing payments whether mailed in, via an electronic medium, by credit or debit card or any other method, and updating payment records. Also included is accepting in-person payments, both through company-operated business offices and third-party pay stations or agencies.

Credit and collections includes any task related to collection of late payments including back office and field activities.

Customers expect utilities to provide many payment options these days. Leading utilities have made technologies such as Web self-service, and interactive voice response (IVR) a focus of customer satisfaction improvement efforts. Energy utility customers now expect to be able to access information regarding accounts, view and pay bills, prepare a home energy audit, and obtain information on energy-saving appliances or home modifications. Energy utilities have also used this expansion of payment options and self-service information as an opportunity to close business offices.

Some water utilities are providing similar services, but the water industry generally lags behind energy utilities in this area. Internet billing and flexible payment options are increasingly common within the water industry. Some water utilities are using IVR to provide information such as bill and payment information and directions and office hours, and many options such as the ability to extend payment due date, enter a meter reading, and make a change to property ownership details. As of now however, IVR adoption is slower than use of the Internet.

The provision of multiple payment options is associated with high levels of customer satisfaction. For example, JD Power and Associates' 2003 Gas Utility Residential Customer Satisfaction Study found that customer satisfaction is highest amongst those who believe that their gas utility performs well in providing pricing and payment options. Other research conducted for Yarra Valley Water (Australia) showed the need for water utilities to keep up with business norms, particularly payment channels generally available, or risk declining customer satisfaction. And similar independent research conducted for the Spartanburg (SC) Water System found that customers want convenient payment options.

Practices vary as to whether to pass on the cost of the more expensive payment options to customers. Many utilities do so; others do not. In the short term, utility operating costs have risen due to the provision of multiple payment options, except where water utilities are passing on the extra cost on to customers as a convenience fee.

Some utilities are also shutting down over the counter payment facilities or considering it, now that there are so many other ways to pay, especially if other in-person payment methods are available at agencies. This is countering, at least partly, the extra costs of accepting credit cards and other more expensive payment options. Others have no plans to shut down their over the counter service, and find other benefits such as pooling counter staff with call center staff and other administrative functions to better match work volume to staffing.

Most participants bill customers monthly (Figure 5.1). However, a few companies bill less frequently, either bi-monthly, quarterly, annually, or on an as-needed basis.

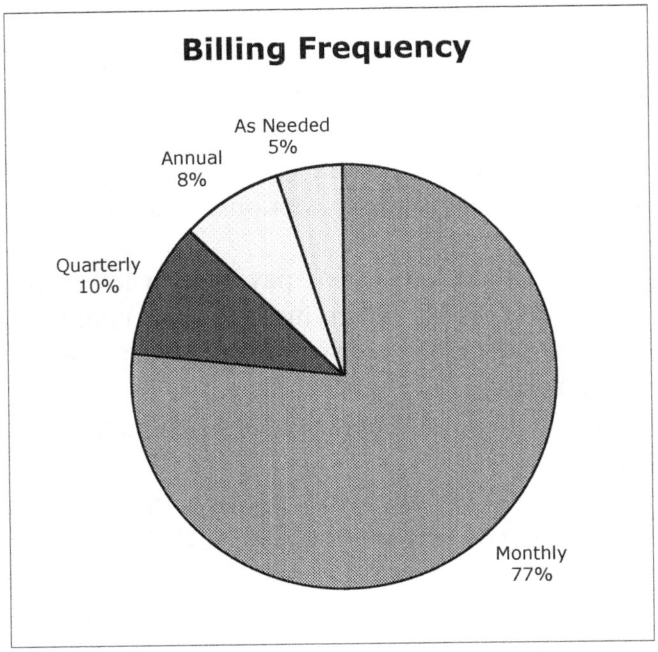

Figure 5.1 Billing frequency

Companies from a range of industries average from 10 to 30 days between bill printing and due date, with utilities near the middle at 17 days (Figure 5.2). This is quite a good result from a cash flow viewpoint, but Telecom/TV companies do even better.

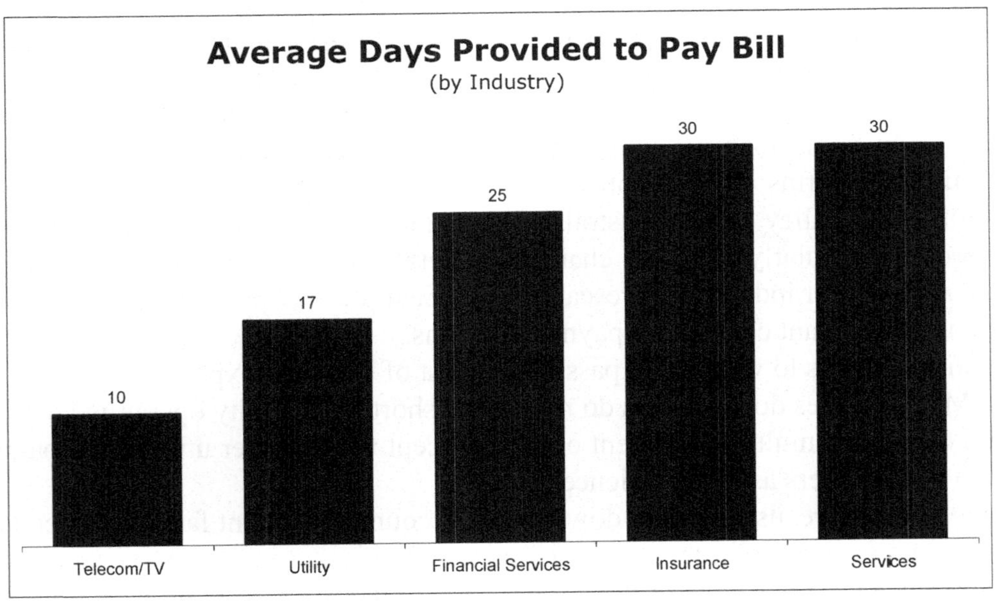

Figure 5.2 Average days provided to pay bill

All participants operate at least one company business office in which customers can come in and pay their bill or transact other business with the utility. Most have just one location, as seen below in Figure 5.3. None of these offices contain a drive-through service.

Only 44 percent of participants also offer other places to pay through a third-party (agency or pay station) network. Agency pay stations are a growing trend in the utility years. Many companies have completely closed company-owned business offices, replacing them with a substantial network of pay station agents. Others have chosen to reduce the number of offices to a minimum yet add more locations through a third-party. Supplementing pay locations with a third-party network can increase customer satisfaction as customers have more locations and can pay during non-business hours.

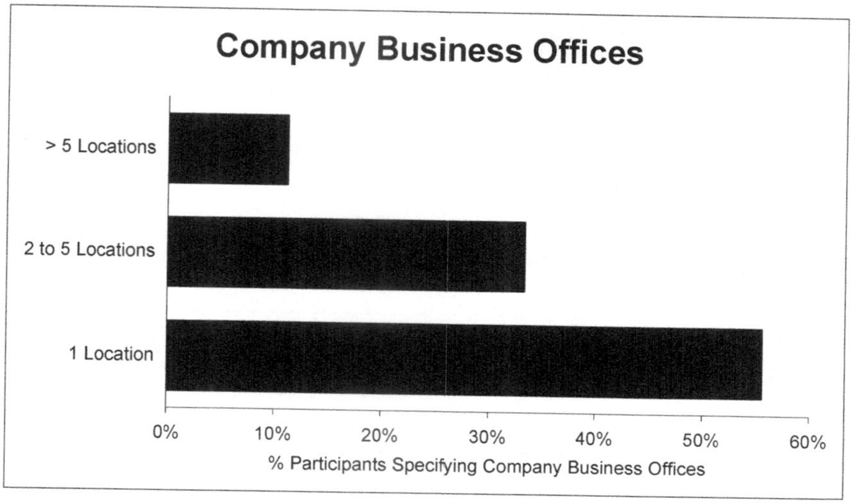

Figure 5.3 Company business offices

As seen in Figure 5.4 below, only 22 percent of participants accept other utility payments at their company offices. Joint payment centers or the capability to accept another utility's payments can open another channel to customers as well as further utilize company assets. Additionally, this is very convenient to customers who prefer to pay in person, allowing them to pay the water bill and the electric or gas bill at the same time, in one stop.

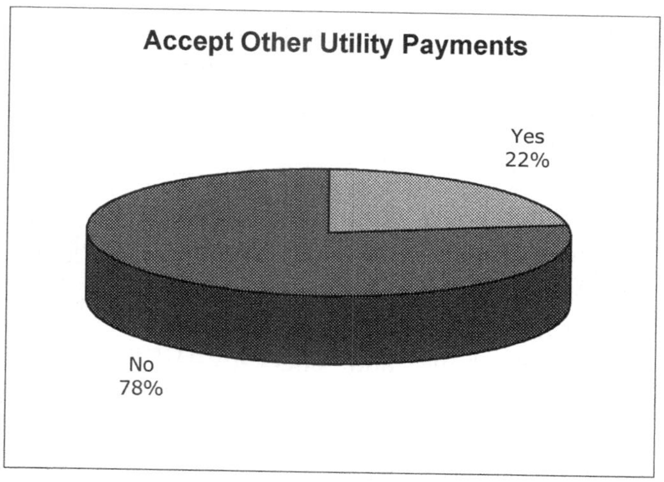

Figure 5.4 Participants that accept other utility payments

Across a range of industries, mail is still the most common channel for bill payment among individual customers (see Figure 5.5). On average, more than two-thirds of customers pay by sending a check in the mail. The next most popular payment channel is through automatic direct debit (otherwise known as Automatic Clearing House or ACH). While direct debit has been offered for 10 to 15 years, many companies are re-emphasizing this as an option to customers.

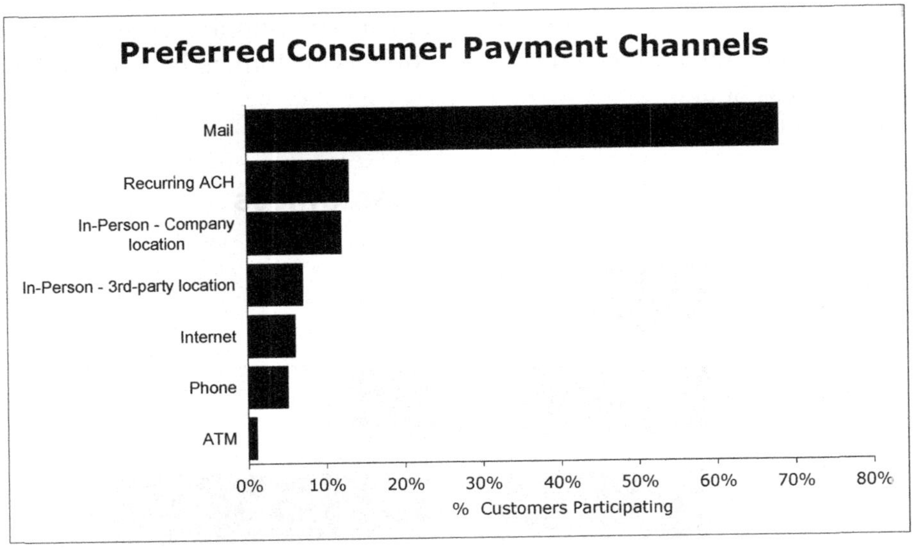

Figure 5.5 Most common consumer payment channels

After paper check, the most common consumer payment method is by debit card—25 percent of consumers pay with a debit card. The next most common method is by credit card—20 percent pay in this manner.

When available, a significant number of customers are still paying bills in person, either at a company location or a third-party location. While many companies have closed walk-in payment locations, primarily because they are the most costly channels to serve, more individual consumers are currently paying through this channel than through the Internet.

Internet payment, pay by phone and other methods are only used by a small percentage of consumers. Internet payment is expected to grow due to the rapid growth in Internet usage by Americans. When asked to identify the last billing or payment option implemented, the most popular offering was Internet Payment (28 percent) followed by Electronic Bill Payment and Presentment (EBPP) (16 percent) and Credit Card Acceptance (16 percent). Clearly electronic payment and credit card acceptance are the key emphasis for companies offering new payment options.

While consumers like debit and credit card payment for convenience, they are more expensive to transact, due to merchant fees. Many companies rely on third parties to handle credit and debit card transactions.

One-time and recurring ACH payments are also becoming more popular. More companies are offering "check-by-phone" and "check-by-IVR". These payment services capture debit information each time the customer chooses to pay the bill over the phone.

Mail is also the most common bill payment channel for business customers (82%), followed by Internet payment (Figure 5.6). Business customers have adopted electronic bill presentation and payment more so than consumers (17 percent versus 6 percent). Long before the

Internet became mainstream, many companies had the capability to send bills and accept payment via an electronic data interchange (EDI) transactions. As a result, the transition to EBPP for many business customers was thus much less of a leap into the unknown.

Ten percent of business customers still pay face-to-face, usually at a company location.

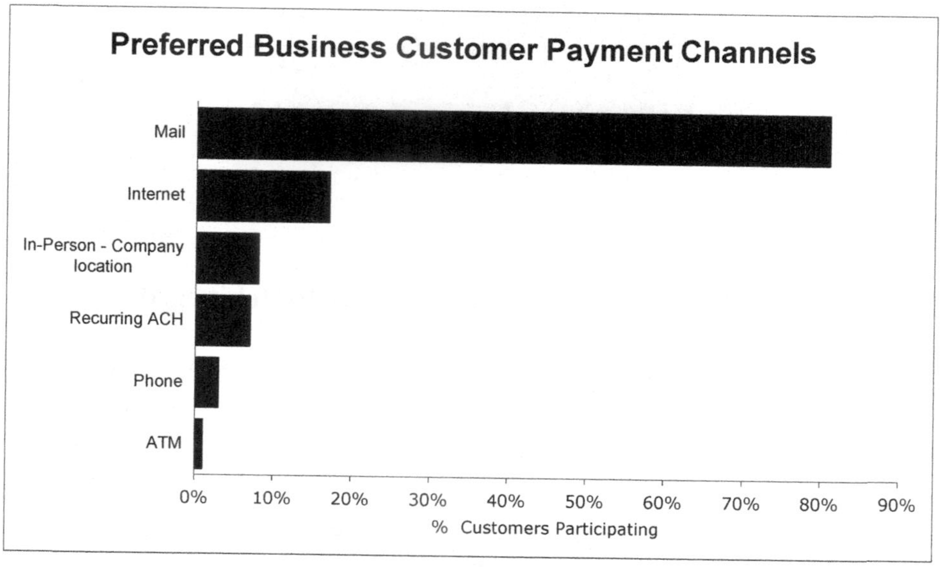

Figure 5.6 Most common business customer payment channels

Nearly 60 percent of participants accept credit or debit card payment—either in person, by phone, by IVR, or over the Internet (see Figures 5.7, 5.8, 5.9 and 5.10). The majority of survey respondents accept credit cards over the Internet (50 percent) or phone (29 percent). About 20 percent of participants accept credit card payments in walk-in locations (in person). No companies accept credit card payments through a kiosk and only one company accepts credit card payments by mail.

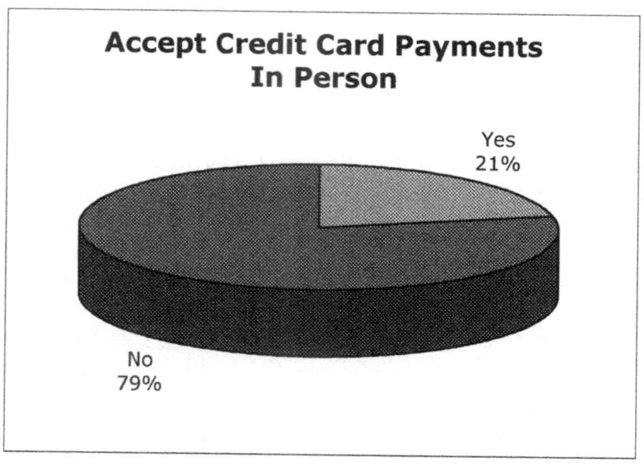

Figure 5.7 Participants that accept credit card payments in person

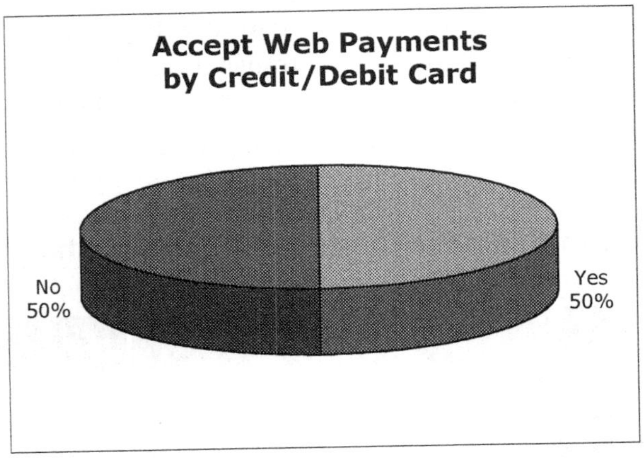

Figure 5.8 Participants that accept web payments by credit/debit card

As shown below in Figure 5.9, less than one-third of participants accept credit card payments over the phone (through a representative). Even fewer companies accept credit card payment through their Integrated Voice Response (IVR) applications (Figure 5.10).

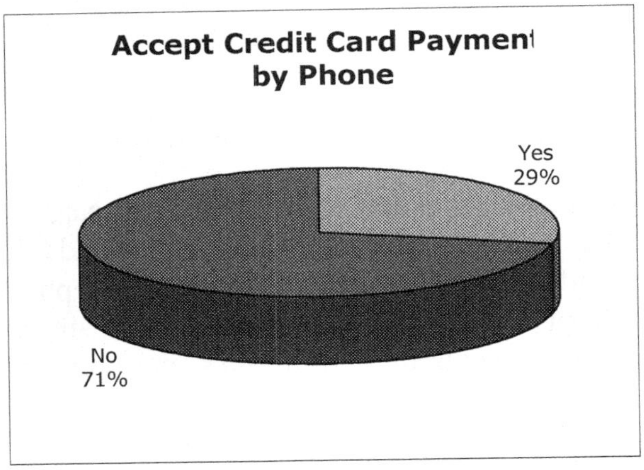

Figure 5.9 Participants that accept credit card payments by phone

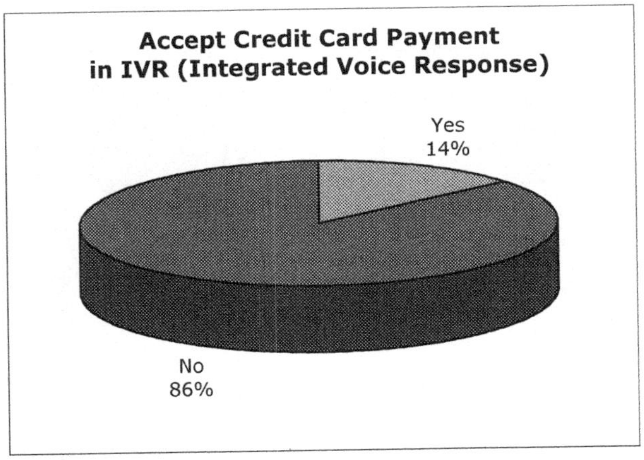

Figure 5.10 Participants that accept credit card payments within the Integrated Voice Response (IVR) system

About one-third of participants charge a fee for a credit/debit card payment (Figure 5.11).

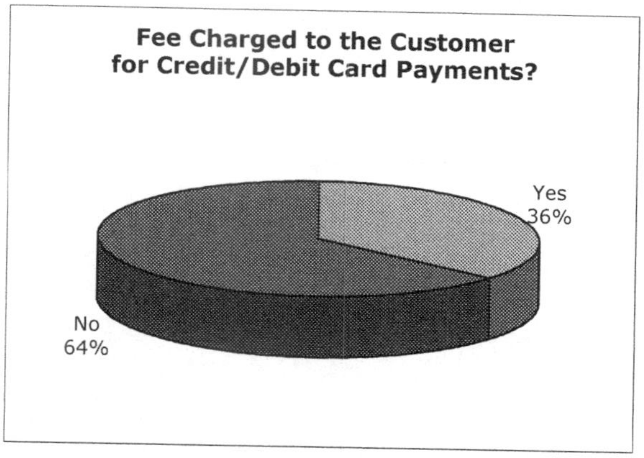

Figure 5.11 Participants that charge a fee to the customer for credit/debit card payments

Only 21 percent charge a fee for check by phone (Figure 5.12) and even fewer, 14 percent, charge a fee for check by web payments (Figure 5.13).

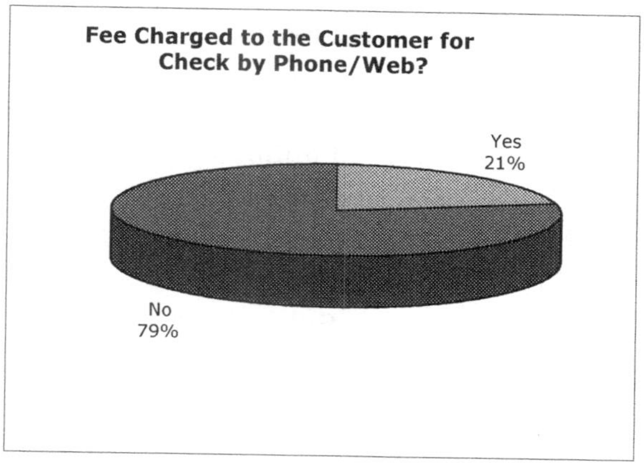

Figure 5.12 Participants that charge a fee to the customer for check by phone/web

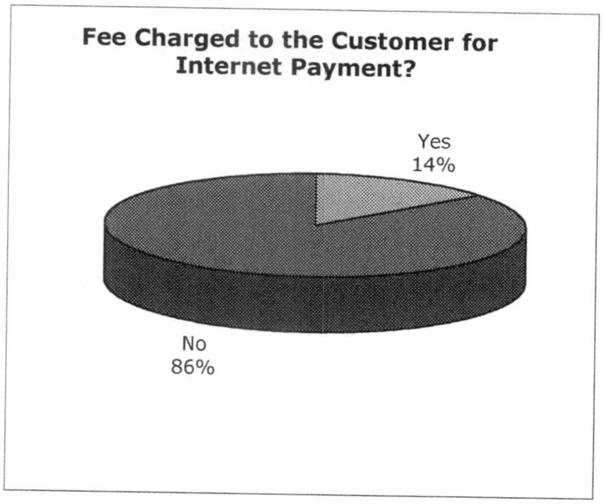

Figure 5.13 Participants that charge a fee to the customer for Internet payment

Less than 20 percent of participants offer prepaid services (Figure 5.14). Prepaid services include allowing customers to pay in a lump sum ahead of time, paying annually, or letting business customers prepay to protect from service interruption or disconnection

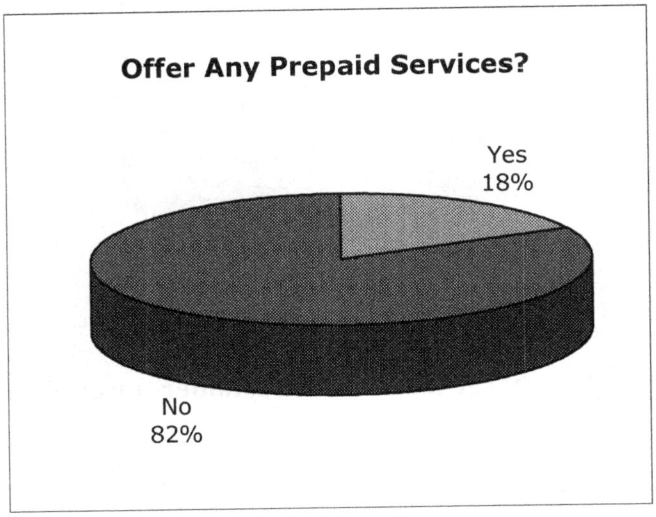

Figure 5.14 Participants that offer prepaid services

As shown in Figure 5.15, the majority of participants do not let customers select their billing due dates, rather customers are assigned billing dates based on the billing cycle and meter reading route.

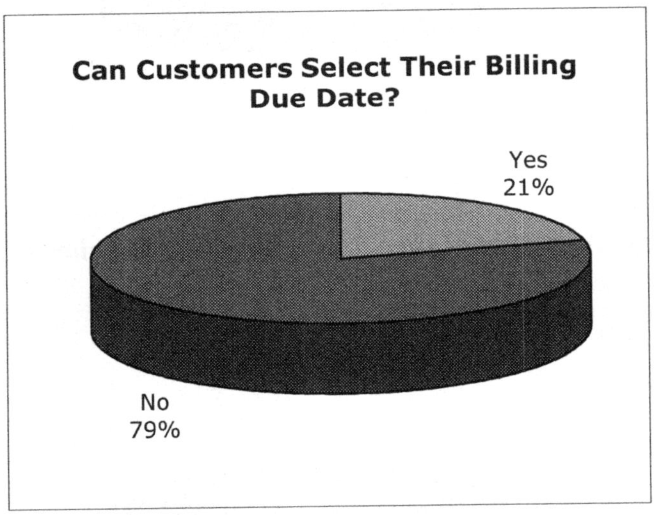

Figure 5.15 Participants that allow customers to select their billing due date

From 7 to 12 percent of consumers are paying their monthly bills in person, as reported by participants (3rd party and company location, respectively).

More than one-third of our participants accept walk-in payments at a company location, primarily municipal utilities and cable companies. The majority of customers paying through a walk-in location are consumers. As seen in Figure 5.16, almost half of our participants accept walk-in payments through a company location and/or a 3rd party pay station. One-fourth of participants accept walk-in payment only through a third-party pay station, usually a grocery store, bank, or check-cashing location.

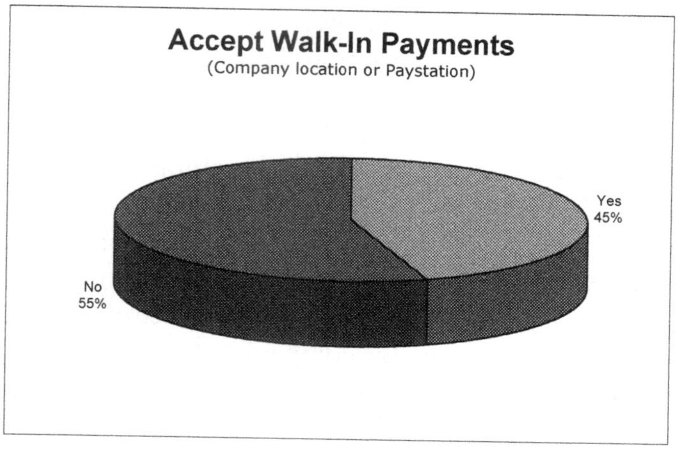

Figure 5.16 Participants that accept walk-in payments (company location or pay station)

It appears that electronic bill presentment and payment (EBPP) is becoming more popular as a payment option for customers. However, as of now, customer adoption except for financial services is quite low. Already, about half of organizations across a range of industries are offering some form of electronic bill presentment and/or payment. Many other companies are implementing EBPP in the near term, or actively considering it.

Only one-third of participating companies have the capability to present bills electronically. Slightly more (38 percent) have the capability to accept electronic payments. But on average, only 7 percent of customers have elected to receive bills electronically (email or web) and fewer (5 percent) pay electronically through the web. Financial Service Firms (Banking, Mortgage, Financing) reported the best adoption rates (Figure 5.17). Very few companies offer any incentives or discounts for EBPP participation.

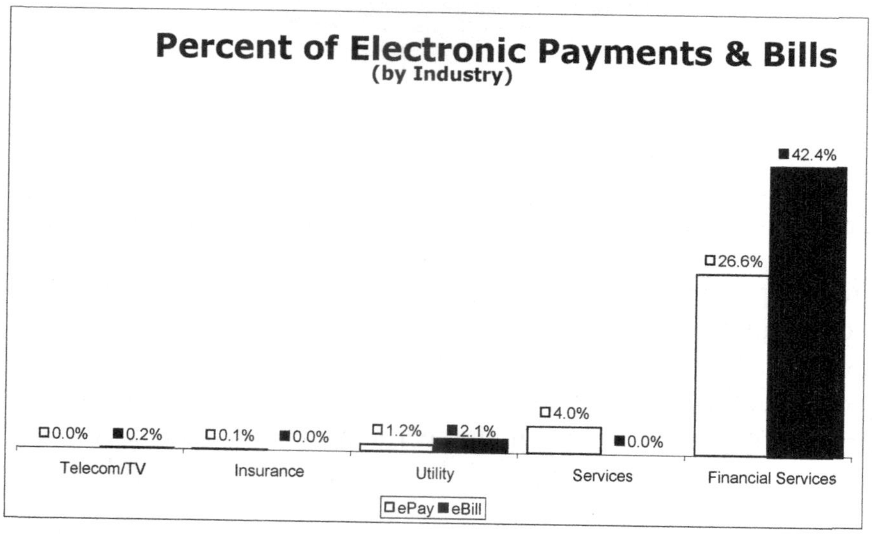

Figure 5.17 Percent of electronic payments & bills, by industry

Utilities reported less than 2 percent electronic bill payment and less than 3 percent electronic bill presentment. J.D. Power and Associates' 2003 Gas Utility Residential Customer Satisfaction Study found that customer satisfaction is highest among those who believe their gas utility performs well in providing pricing and payment options. While 75% of consumers pay their gas utility bill through the mail or by visiting their utility, 90 % of customers would prefer to pay through an alternative method, such as the Internet, payment agency or automatic deduction from their bank account. The study also finds that consumers are open to managing bill amounts through energy efficiency and rebate programs, which may keep potential rate increases from impacting their utility bills.

EBPP also offers significant opportunities to reduce costs, streamline payment processing, and remove paper from payment processing. While electronic payment does offer some savings and the elimination of many manual processes, it's really electronic bill presentation that offers the biggest savings for companies. Other electronic payments initiated by phone, in person, or through an IVR (Interactive Voice Response) unit also offer cost savings.

The percentage of participating water utilities currently offering EBPP is low, with 4 out of 13 water utilities, as compared to an overall participation rate of greater than 50 percent (Figure 5.18). If the above research on gas utilities is any guide, water utilities could see a marked improvement in customer satisfaction by promoting EBPP.

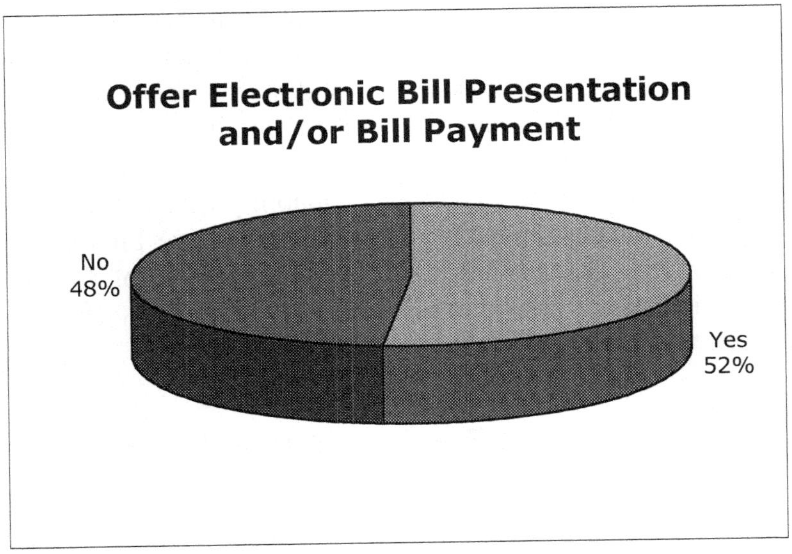

Figure 5.18 Participants that offer electronic bill presentation and/or bill payment

There are two primary models for delivering Internet bills to consumers, the biller-direct model and the consolidator model. Under the biller-direct model, organizations offer payment facilities via their web sites. Under the consolidator model, a third party provides this facility on behalf of many organizations. Consolidators also offer the option to set-up recurring payments to service providers, thereby encouraging a higher participation rate among consumers.

As Figure 5.19 below shows, there is about an even split currently between each model, with a substantial number doing both.

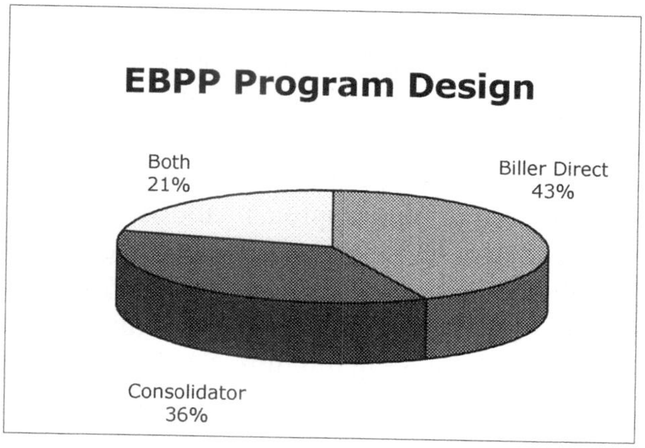

Figure 5.19 EBPP program design

Economic downturn has lead to an increase in uncollectibles in all industries. Rising fuel costs have further impacted uncollectibles for energy utilities, which have responded by focusing greater efforts on credit and collection activities:

- Improved identification, segmentation, and prioritization of customers, to identify credit risk earlier and treat appropriately:
  - Sharing credit and payment history with other utilities
  - Credit scoring and behavioral analysis
  - Fraud detection analysis

- Technology and automation:
  - Mobile data terminals for dispatch and execution of collection action
  - Outbound dialers to remind customer of past due and/or collect
  - Better integration between credit/collection systems and billing system

As Figure 5.20 below shows, half of participants are required to observe a moratorium on collection actions during certain times of the year. Moratoriums generally prohibit utilities from disconnecting service to customers who fail to make payments, usually during winter or summer months to protect customers from loosing utility service.

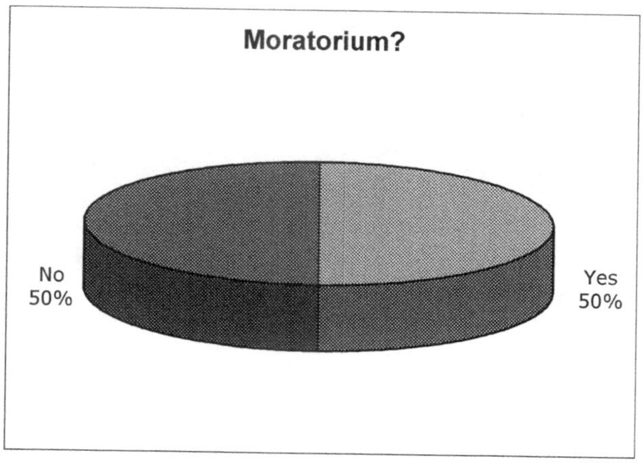

Figure 5.20 Use of moratoriums by participants

Less than one-fourth of participants use credit scoring in their credit and collection policy (Figure 5.21). Credit scoring and behavioral scoring models can be used effectively to segment and analyze customer behavior, predict risk and reduce collection costs.

Credit scoring on the front-end allows the appropriate deposit decisions to be made. One company in our study is using credit scoring to determine the need for a deposit (for residential customers). When regulations permit, this approach allows good paying customers to avoid paying a deposit, while poor risk customers are required to place a deposit.

Behavioral scoring helps minimize risk of default as well as minimizing cost of collection. Knowing the payment habits and behaviors of your customers will help you determine the collection actions and timing that will be the most effective.

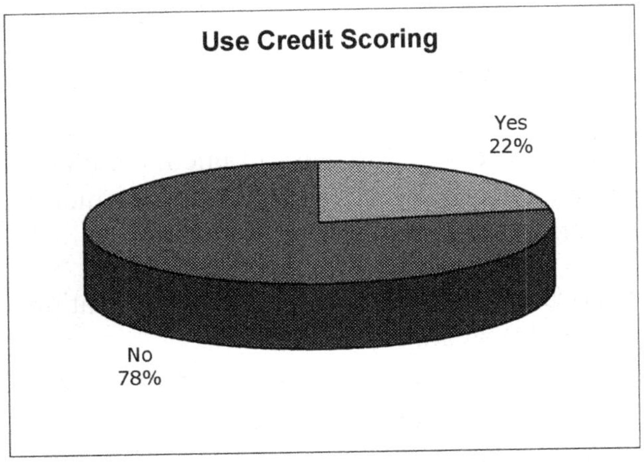

Figure 5.21 Use of credit scoring by participants

As Figure 5.22 below shows, only one-third of participants exchange bad debt information with other utilities or exchange bureaus. Sharing bad debt information can be an effective skip-tracing technique as customers move from one service-territory to another.

A little more than half (56 percent) of participants report derogatories to credit bureaus. Reporting payment history to credit bureaus can be an effective motivation for chronically slow or late paying customers. Customers are more likely to pay creditors that do report a derogatory before paying those creditors that don't, due to the potential impact on the customer's credit file.

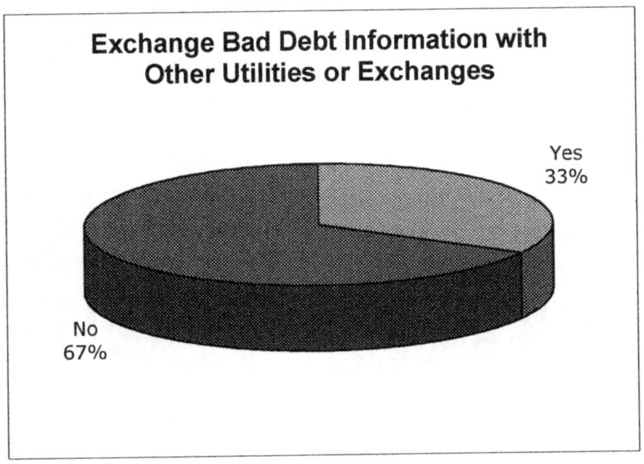

Figure 5.22 Participants that exchange bad debt information with other utilities or exchanges

Two-thirds of participants are utilizing automated dialers or IVRs to automate outbound dialing and late payment notification (Figure 5.23). Predictive dialers especially can be a very cost effective collection technique and are the mainstay for other industries (financial services, telephone, cable). Outbound dialers can also be used effectively to notify customers of an impending cut-off or to remind customers of a late payment, a much cheaper alternative than a trip to the field.

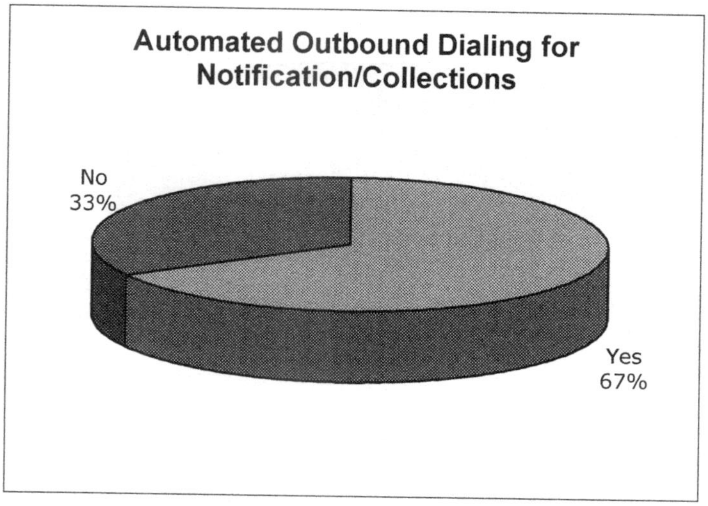

Figure 5.23 Use of automated outbound dialing for notification/collection by participants

Nearly half of participants require or can ask for a deposit, for residential customers (Figure 5.24). Deposits collected from customers average about a 2-month bill. Collecting deposits is a good way to minimize losses. Having a strong, consistent deposit policy is a best practice in the utility industry. Using credit scoring to determine the need for a deposit further strengthens the policy by rewarding good credit history customers (with no deposit) while protecting the company from bad debt.

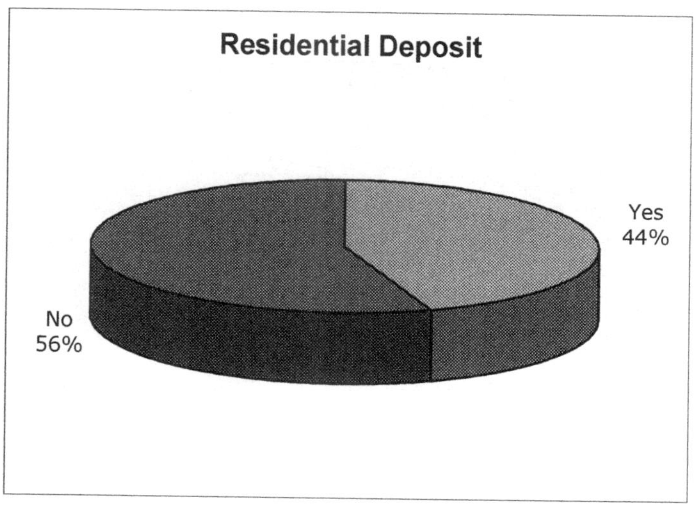

Figure 5.24 Participants that require a residential deposit

More than half of participants outsource some collection activity (Figure 5.25). All those outsourcing in this study have outsourced finaled or inactive accounts to one or more collection agencies. Outside agencies can be very effective at skip tracing, handling bankruptcies, and chasing inactive accounts. On average, those outsourcing inactive collections in this study have been doing so for more than 5 years.

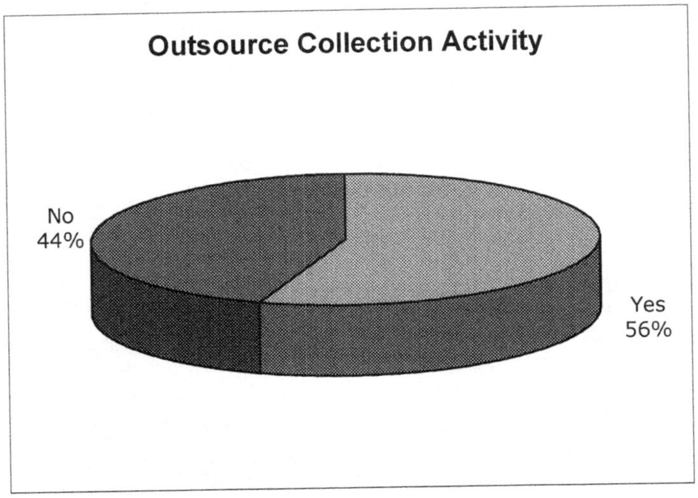

Figure 5.25 Participants that outsource collection activity

Only one-third of participants reported using a dedicated credit and collection management system (Figure 5.26). The remainder relies on the customer information system to prioritize collection work and manage receivables.

Specialized credit and collection systems can provide powerful tools for better management of receivables, better prioritization of collection action, and better tracking of customer behavior.

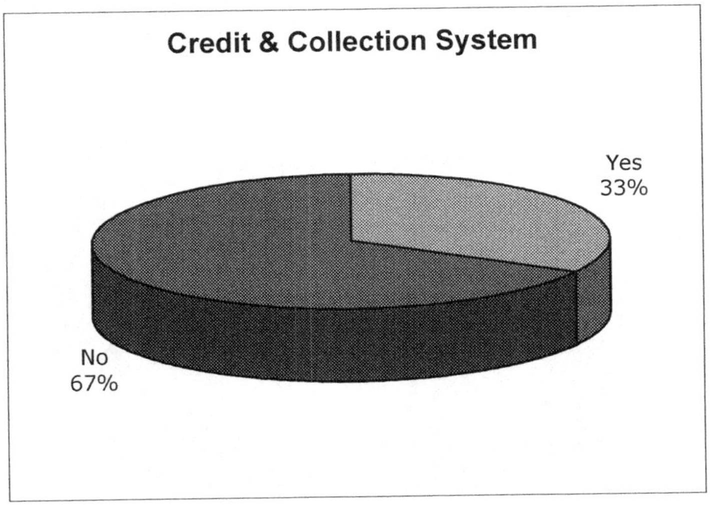

Figure 5.26 Credit & collection system.

METRICS AND BEST PRACTICES

Billing & Payment

Key efficiency and effectiveness metrics were calculated from data supplied by approximately 15 utilities, including 10 water utilities. Table 5.1 below contrasts the best performers (those utilities that had both above average efficiency and effectiveness) with the utility industry and participating water utilities.

The key billing metrics we have chosen to provide a balance between service and efficiency are:

- Bills per involved FTE (annual)
- % Rebills (annual) (bills that are re-issued or adjusted as a result of a customer initiated contact or internal correction)
- % Unbilled Accounts (accounts not billed during a representative month)
- % Unbilled Revenue (of total revenue, for a representative month)
- Unit cost (cost per bill – Operational & Maintenance costs only – direct labor, contractor costs, overtime, and non-labor O&M; no capital costs or overheads)

To maintain a balanced view of performance, utilities were designated as "best performers" if they exhibit above average performance in both cost and service. Once the "best performers" were identified, "best performer average" metrics were calculated, as reflected in Table 5.1 below. In the sample, there were 2 "best performers" identified out of 15 utilities participating, of which 10 were water/waste water utilities.

Table 5.1 Billing and payment benchmarking performance

Billing & Payment Benchmarking Performance			
All Utilities	Best Performers	All Utilities	Water Utilities
Bills Per FTE	928,651	409,357	331,035
Percent Rebills	1.14%	1.97%	3.03%
Percent Unbilled Accounts	0.01%	0.39%	0.08%
Percent Unbilled Revenue	0.2%	0.1%	0.0%
Cost per Bill	$ 0.29	$ 0.29	$ 0.42
Utilities Participating	2	15	10

Remittance Processing

Key efficiency and effectiveness metrics were calculated from data supplied by approximately 30 utilities, including 8 water utilities. Table 5.2 below contrasts the best performers (those utilities that had both above average efficiency and effectiveness) with the utility industry and participating water utilities.

The key metrics chosen to provide a balance between service and efficiency are:

- Payments processed per FTE involved per month
- % errors (for a representative month)
- % carry over (payments not processed on the day of receipt)
- % same day cash (funds that can be used the day they are collected)
- Cost per payment processed – direct labor, contractor costs, overtime, and non-labor O&M; no capital costs, benefits or overheads)

To maintain a balanced view of performance, utilities were designated as "best performers" if they exhibit above average performance in both cost and service. Once the "best performers" were identified, "best performer average" metrics were calculated, as reflected in Table 5.2 below. In the sample, there were 4 "best performers" identified out of 30 utilities participating, of which 8 were water/waste water utilities.

Table 5.2 Remittance processing benchmarking performance

Remittance Processing Benchmarking Performance			
Remittance Processing Benchmarks	Best Performer Average	Utility Average	Water Industry Average
Payments Per FTE	1,423,625	621,253	320,938
Percent Errors	0.03%	0.44%	0.12%
Percent Carry-Over	3%	13%	22%
Percent Same-Day Cash	93.8%	77.5%	83.9%
Cost per Payment	$ 0.12	$ 0.32	$ 0.36
Utilities Participating	4	30	8

Credit & Collection

Key efficiency and effectiveness metrics were calculated from data supplied by 9 utilities, including 6 water utilities. Table 5.3 below contrasts the best performers (those utilities that had both above average efficiency and effectiveness) with the utility industry and participating water utilities.

The key metrics chosen to provide a balance between service and efficiency are:

- Delinquent transactions per FTE involved (per month)
- Cost per delinquent transaction processed – direct labor, contractor costs, overtime, and non-labor O&M; no capital costs, benefits or overheads)
- Net write-offs (% of total revenue)
- ENDRO (Equivalent number of days revenue outstanding)
- Cost of credit (=((Write-Offs + Carrying Costs of Receivables + O&M Expenses + Unbilled Revenue) - (Service Charges + Net Interest on Deposits)) / Total Revenues)

To maintain a balanced view of performance, utilities were designated as "best performers" if they exhibit above average performance in both cost and service. Once the "best performers" were identified, "best performer average" metrics were calculated, as reflected in Table 5.3 below. In the sample, there were 3 "best performers" identified out of 9 utilities participating, of which 6 were water/waste water utilities.

Table 5.3 Credit & collection benchmarking performance

Credit & Collections Benchmarking Performance			
All Utilities	**Best Performer Average**	**Utility Industry Average**	**Water Industry Average**
Average of Delinquent Transactions Per FTE	6,440	3,835	3,894
Average of Cost Per Delinquent Transaction	$ 25.32	$ 217.72	$ 338.42
Average of Net Write Offs	0.36%	0.83%	1.10%
Average of ENDRO	21.9	34.0	30.3
Average of Cost of Credit	$ 0.49	$ 0.74	$ 0.74
Utilities Participating	3	9	6

Business Offices

Key efficiency and effectiveness metrics were calculated from data supplied by 10 utilities, including 9 water utilities. Table 5.4 below contrasts the best performers (those utilities that had both above average efficiency and effectiveness) with the utility industry and participating water utilities.

The key metrics chosen to provide a balance between service and efficiency are:

- Walk-in transactions per FTE involved (annual)
- Average wait time (minutes)
- Cost per walk-in transaction processed – direct labor, contractor costs, overtime, and non-labor O&M; no capital costs, benefits or overheads)

To maintain a balanced view of performance, utilities were designated as "best performers" if they exhibit above average performance in both cost and service. Once the "best performers" were identified, "best performer average" metrics were calculated, as reflected in Table 5.4 below. In the sample, there was 1 "best performer" identified out of 10 utilities participating, of which 9 were water/waste water utilities.

Table 5.4 Business office benchmarking performance

Business Office Benchmarking Performance			
All Utilities	**Best Performer Average**	**Utility Industry Average**	**Water Industry Average**
Walk-in Transactions per FTE	53,230	15,621	15,621
Average Wait Time	1.00	2.75	2.75
Cost per Walk-in Transaction	$ 0.94	$ 5.69	$ 5.69
Utilities Participating	1	10	9

CHARACTERISTICS OF BEST PERFORMERS

Best performing companies are offering customers many options for billing and payment to make it easier for all customers to receive and pay their bills. This includes opening all channels to payments—in person, by phone, by IVR, on the Internet, and by mail. High performing companies are encouraging and promoting low cost payment channels to maximize efficiencies and reduce overall operating costs. More specifically, best performing companies:

- *Continually research customer billing and payment preferences.* Customer spending patterns change over time and with the introduction of new technology. High performing companies are conducting periodic customer research to better understand customer billing and payment preferences, by customer segment. As demonstrated in this study, consumer and business payment preference vary significantly. Customer research will arm you with the information you need to match service offerings to preferences.

- *Accept debit and credit cards.* Aside from paper checks, the debit card was the most popular method of payment for consumers, followed closely by credit card payments. Debit cards are readily accepted in the retail industry and consumers are becoming accustomed to swiping a debit card in lieu of writing a check. Additionally, loyalty programs have increased credit card spending. Expanding payment options to accept debit and credit cards makes it more convenient for customers to pay.

- *Extend direct debit payment options.* Most customers pay their monthly bill with a check. Our best performers are introducing payment options to accept check-by-phone, check-by-IVR, and check-by-web (non-recurring direct debit or ACH payments) to make it easier for customers to quickly pay their bill. Non-recurring direct debit payments can be easily automated to eliminate handling and lower transaction costs.

- *Promote electronic billing and payment options.* Best performers make sure that customers are aware of all electronic billing and payment options to increase acceptance and usage. Traditional and non-traditional advertising and promotion can be used to introduce options and expand adoption. Print reminders on the bill, on the insert, as well as on the return envelope. Train customer service representatives to refer customers to electronic billing and payment services. Ask new customers for email addresses during service initiation, and if they'd like to receive electronic billing statements. Do everything possible to make sure customers know about electronic payment options and find it easy to make a payment.

- *Offer incentives to encourage EBPP participation.* Incentives and discounts do increase participation. Make sure the incentives are fair and worth the effort. Conversely, transaction fees discourage participation. Make sure your fees and discounts are pointing customers to the lowest cost channels.

- *Provide agency payment facilities, such as banks or Western Union, to increase customer convenience and satisfaction.* Contracting with a network of third party paystations makes it easier for customers to pay their bills—by adding more locations and extending your hours of operation.

- *Use credit information to segment and rank collection work.* High performing companies using credit scoring and behavioral scoring models to segment and analyze customer behavior as well as predict risk. Credit scoring on the front-end allows the appropriate deposit decisions to be made. Behavioral scoring helps minimize risk of default as well as minimizing cost of collection. Credit history is a useful tool, but most companies do not use it effectively. Best performers gather and maintain critical information about their customers to help identify risk and to determine appropriate collection action. Not all customers should be treated the same, knowing the payment habits and behaviors of your customers will help you determine the collection actions and timing that will be the most effective. Segmenting and prioritizing accounts will also reduce collection costs.

- *Have a consistent credit and collection policy and do what they say they'll do.* Best performers act on a timely basis. These companies consistently execute disconnect policy to demonstrate intention to chronically delinquent customers—they follow their policy and do what they say they'll do. Consistency in policy will encourage the appropriate behaviors. Customers become conditioned to expect what they see in action, not what they are told. Most utilities deliver many more cut-off notices than they work. This sends the wrong message to customers. In addition, inconsistency in policy execution leads to complaints.

- *Train all employees to act consistently.* Best performing companies have trained both field and call center employees to treat customers consistently and to follow policies and procedures. Employees need to respond consistently to discourage customers from "shopping" for extensions and special treatment. Consistency between field and office employees also delivers a consistent message to customers. Additionally, best performing companies use automation as much as possible to support employees in the consistent delivery of collection treatment.

- *Automate as much as possible – extension tracking, letter generation, dunning, matching, and prioritizing workload.* Best performing companies take advantage of automation to deliver letters on a timely basis, track payment commitments, match uncollectible balances to new customers, and prioritize collection actions. Collection systems help companies manage accounts receivables, including tracking accounts, prioritizing tasks based on aging, value, risk, and collectibility.

- *Utilize IVRs or predictive dialing equipment to remind late payers and to warn delinquent customers of impending cut-off.* High performing companies take advantage of computer-telephony to contact customers, both as a courtesy and as notification of cut-off. Automation reduces the operational cost of these calls and enables more calls to be made, even after normal business hours. Companies have found that many "late payers" appreciate a courtesy call as a reminder. Notification of cut-off by phone is a low-cost way to offer customers one last chance to avoid disconnection, and much cheaper than a field visit.

- *Network with credit bureaus and other utilities to identify problem customers, recover unpaid finals.* A small percentage of customers will work the system, using service for as long as possible before being disconnected for non-payment. Often the service is put in someone else's name to continue getting service or the customer will move and repeat the process. Best performers network with other utilities and credit bureaus to identify these transient delinquent customers and require payment prior to service (depending upon regulations). Networking and sharing collection history helps utilities identify high-risk customers so they can be managed appropriately at the same time as reducing the risk to the company.

- *Work to the rules.* Best performers work closely to their credit and collection policies and state and federal regulations. Most companies do not do all they are allowed by their regulators, tending not to act soon enough, primarily as a way of avoiding customer complaints. Companies extend due dates much longer than required, are lenient on collecting deposits, and do not follow-up with poor paying customers. In addition, reviews of regulations are only carried out occasionally. High performing companies work closely to the rules, especially for high-risk customers. Segmentation of the accounts lets companies treat the right customers with the most effective collection actions, at the most effective time.

CASE STUDIES

Kansas City Power & Light (KCPL)

KCPL provides electricity to over 490,000 customers in 24 northwestern Missouri and Kansas counties, a territory of about 4,600 square miles. KCPL customer research has revealed that billing-related notices and transaction confirmations in general are very important to customers. As a result, KCPL offers email notification when a customer bill is issued and a reminder email 5 days prior to the due date (the reminder is not sent if the customer has already paid the bill). KCPL automatically emails customers if there is a problem with the submitted banking information. KCPL sends a confirmation email if a customer's email address is modified.

If an email to a customer comes back as undeliverable, KCPL flags it as "bad" in its database, causing a message to pop up when the customer logs in notifying that the customer needs to update his/her email address on file. This enables KCPL to maintain a very reliable email list, and it makes it convenient for the user to maintain and make changes. About two-thirds of KCPL's customers who register for online account access also register for online bill payment, and nearly all online bill payment customers request billing notification via email.

KCPL removes dormant electronic accounts when there has been no access for 12 months. A reminder is sent at 360 days warning of pending deactivation, which can be averted by logging in from the link attached in the email. If the customer passes on this, the accounts are removed from the online system. However, KCPL also learned that some customers were receiving email notices and using those as reminders to mail in their check. When KCPL removed them from its online system, they stopped receiving email notices (and were not happy!). Now, when a customer chooses to receive email notices, KCPL exempts them from the deactivation rule.

KCPL has also implemented special electronic payment options for business customers since many do not permit standard debit transactions on their bank accounts. These include (1) Web Approve--customer approves bills for payment on the Web site, then initiates an ACH credit (deposit) into KCPL's bank account; (2) ACH Push--customer sends KCPL an ACH credit payment through an EDI 820 format containing account level remittance data. Business customers like both of these options because they allow deposit into a KCPL bank account which means they do not need to open their bank account to a vendor for withdrawals. It also means that combined with online bill presentment they are getting same-day bill availability and same-day payment recognition, which helps on tight A/P turnarounds.

KCPL has made its credit/debit enrollment process logical and intuitive, making features optional wherever possible. Some customers do not want email notices, while others do not want to give up the paper invoice. KCPL asks new registrants enrolling into online payment if they would like email bill notification. It also asks if the customer still wants the paper bill. As a result, even though KCPL offers no incentive, 40% of customers select the option to stop receiving a paper bill in the mail.

In 1994 KCPL also invested in a wireless meter-reading system (Cellnet) that gave utility employees instant access to meter readings. KCPL now uses this system to offer Account Link, which enables customers with Internet access to monitor their power usage minute by minute.

Fannie Mae

Fannie Mae is a government-backed listed mortgage underwriter and one of the largest financial institutions in the U.S. Fannie Mae used an ROI model to determine best EBPP implementation model. It also tracked actual ROI and made adjustments to its EBPP rollout strategy to achieve ROI targets. Not all options were implemented at once. Some EBPP features made ROI sense only after a certain level of customer adoption had been achieved. Fannie Mae used its EBPP platform to eliminate redundant manual reporting systems, thereby achieving greater cost reduction and efficiency.

Fannie Mae provides only single page paper invoicing. In order for customers to receive their billing detail, they must get it from Fannie Mae's online bill presentment platform. Only a one-page bill summary is sent through the mail for customers who have not moved to paperless invoicing. Fannie Mae used Single Page Invoicing (SPI) to encourage EBPP adoption, finding it much easier to convince customers to use SPI when compared to paperless invoicing. Also, once customers are accustomed to SPI and Electronic Bill Payment, it is much easier to convince customers to go paperless.

Fannie Mae made its EBPP system available to its customer service/sales reps so they could work more efficiently with customers. In due course, representatives had a better understanding of the system and encouraged a higher customer adoption rate of the service.

Yarra Valley Water

YVW is a state government-owned Australian water/wastewater utility serving 680,000 accounts in Melbourne, Australia's second largest city. YVW's payment channel strategy has historically focused on customer satisfaction with the secondary goal of efficiency. Consequently it now has a diverse range of payment channels (e.g. a customer can pay by phone using a credit card via the IVR, at post offices and over the Internet).

Daily, electronic file uploads from each payment channel (credit and debit cards, checks, in person, etc) occur with no human intervention into a common internal system. The majority of payments are made in person at post offices, but banks are starting to compete for this business and are winning it away from the post office. YVW does not offer in-person payment at its own offices; this is totally outsourced on a competitive basis. As with most customer service functions in Australia, there is a very competitive market for the outsourced over the counter payment business. Many Australians pay almost all their bills this way.

YVW has a very small reception area, preferring to offer its services to customers through a range of channels and partners. These partners, such as post offices and banks, have service centers distributed across the service area. Bill payment can be made in person via any post office. In 1995, there were up to 35 employees in this area; now there are 2.

Applications for connection and plans can be made through a franchised plumbing supplier (similar to a mini Home Depot), via YVW's "EasyAccess" system, offering automated response

(continued)

Yarra Valley Water (continued)

and approval for standard connection applications. The EasyAccess system was designed in house, and has won International Technology Awards and recognition from the Smithsonian Institute in Washington, DC. The system provides real-time interfaces between the plumbing supplier and the utility, with 75-80% of applications approved based on programmed business rules without human intervention. The approvals are faxed to the plumbing supplier, usually within 15 minutes. The applicant typically submits the application, buys his/her materials, then picks up the faxed approval and all needed documentation and plans at the checkout.

Future plans include web self-service to eliminate the need for a front office.

Payment processing errors are very low at 0.09%, and are mainly the result of debit payment attempts when there are insufficient funds in the customers' accounts.

YVW currently outsources the majority of its credit management functions to an external debt collection agency. The reason for the contracting out of debt collection was the inability to contact sufficient debtors when compared to the flexibility of resources offered by an outside agency.

The debt collection focus with YVW's contractor is on both efficiency in terms of expected recovery rates, and customer service, with a number of performance measures established for a range of key measures. The contract differs from standard debt collection contracts as it focuses more on customer service with key measures around complaints management and the support and referral of customers in financial hardship. The contract also provides a financial incentive, namely a share of the revenue collected as long as the bad debt recovery is more than 70%.

Over the past three years YVW has achieved significant reduction in its long-term debtors, whilst costs have been reduced through a focus on continuous improvement through process re-engineering and leveraging the experience of the external debt collection agency. The utility recruited a specialist credit and collections manager with a track record of turnarounds in this function. Subsequently, many low value activities were eliminated and costs of the function reduced. Once the function was working well, it was outsourced and this contract is re-bid every 3 years.

All utility customers are placed into designated arrears paths based on their customer type (e.g. residential, concessional, pensioner, key account etc) and different credit management approaches are conducted based on the type of customer segment.

YVW's board receives a monthly debtors report to emphasize the importance of this function to the utility's financial health.

Under its Customer Contract, YVW must offer the following payment options and may, at its discretion, offer additional methods: in person at a network of agencies or payment outlets; by mail; by automatic deduction from a savings, check or credit card account; by phone; and by internet. In all, 25 payment channels are offered (11 different credit cards are accepted).

(continued)

Yarra Valley Water (continued)

The key trends for YVW's payment channels over the past two years are:

- Over the counter cash payments have decreased by 11%, but remain the most popular payment instrument at 34% of total payments.
- Credit card payment volumes have grown by 7% and now represent 25% of the total payment volume, which is the second most popular payment instrument.
- Debit Cards are the fourth most popular payment instrument with their usage growing by 41%, to 13% of total payments.
- Direct debit/credit has grown steadily by 5% to 5.7% of total payments. The relatively low volumes are in line with other utilities
- Electronic payments have grown by 11% and as a percentage of total payments have risen to 7%.

A recent detailed payment channel review was completed to identify the opportunities to reduce payment channel costs without significantly impacting customer satisfaction. The review concluded that the approach should be guided by the following principles:

- YVW is unable to significantly influence wider customer payment trends,
- For channels that are growing or constantly popular, the strategy should be to focus on efficiency and if the transaction cost is relatively low, promotion of that channel,
- For channels that are decreasing in popularity, the strategy should be to essentially do nothing. YVW has limited influence in customer trends that are driven by external factors and therefore resources are better directed towards increasing the efficiency of growing channels.

Tucson Water

Tucson Water provides drinking water to the City of Tucson, which has a population of 800,000. Unusually high monthly usages are identified and reviewed based on consumption test parameters before they are billed. This helps reduce billing inquiries. The bill includes a graph of the customer's water usage for 13 months. The graph allows comparison of current to previous usage and depicts the seasonal demands. Customers who want to dispute their bill have the right to an administrative hearing. The hearing officer makes the final decision regarding such disputes.

Staffing in the billing area has remained constant over the past 10 years despite high growth in the number of customers. This has been accomplished by a variety of productivity improvements, such as cross training. Daily work assignments are made based on the work to be done and employee availability, not on fixed job specializations. In addition, planned job rotation keeps skills current and adds to job variety.

Another source of efficiency is multiple service billing; water, sewer and trash collection are all billed on the one bill in Tucson.

Direct debit has grown to 12% of accounts, which is also reducing billing costs. On the other hand, the utility has recently contracted with Western Union to accept customer payments, which at $1 per payment is increasing costs. The local electric company passes on this fee to the customer but Tucson Water has decided not to do so.

As an incentive to minimize past due accounts and recover costs, the delinquency process includes a $20.00 fee for an initial past due notice and a $30.00 fee if the account is locked for non-payment.

Passaic Valley Water Commission (PVWC)

PVWC provides drinking water to more than 750,000 people in northeast New Jersey. By far the most common way that PVWC receives payment is via the mail. The utility provides customers with an addressed return envelope. PVWC as yet does not offer debit or credit card payment (these are coming soon). PVWC uses a lock box service provider who picks up from the utility's PO Box twice a day and deposits payments at its bank, also twice each day.

After processing, files are electronically transmitted and are uploaded to the utility's CIS every day. After the evening update all payments are posted to the customer accounts. The payment processing and file transfer is a very low cost service, at 12 cents per transaction.

PVWC only receives about 130 walk-in payments per day, which are all accepted at its office. In addition, any remittances that cannot be processed are sent via courier each morning to PVWC's cashier for processing. Drop box and walk in payments are processed by a cashier. Deposits are made daily and payments are marked as pending on the day of receipt and posted to the account the next day. All documents, checks and stubs are scanned.

The error rate is not as low as it could be, at about 1%, due largely to the significant number of customers who own more than one property and pay with one check for them all. Since the checks are machine-sorted, if *all* the account numbers are not on the check, these are bounced out, or sometimes the entire amount is credited to just one property.

PVWC also provides water and sewer billing and collection services for several other municipalities. This provides a more cost effective billing and collection solution for others while increasing PVWC's revenue.

Electronic bill presentment using HTE's Click2Gov will be available shortly. PVWC is also currently developing a new interactive web site allowing customers to conduct all business processes online.

PVWC has only 1 person handling over the counter services. The utility serves 68,000 accounts. This person is very experienced, at over 15 years service.

The supply of return envelopes with bills greatly reduced the amount of walk-in traffic. Many other services apart from bill payment are available and are handled by the one person.

PVWC has in recent years adopted a more aggressive collections approach, and this has markedly improved the performance of this function. Customers are now only allowed one payment extension (ever) rather than the previous practice of dealing with customers who were often delinquent in their payments.

PVWC bills quarterly. Bills unpaid after 30 days receive a delinquent notice, which is considered the second notice. On day 60 a door tag is placed, and 7 days later if the bill is still unpaid the water is turned off. There is a $50 service charge for reconnection. Consequently, there are typically only about 60 residential accounts more than 90 days overdue out of 13,000

(continued)

Passaic Valley Water Commission (continued)

residential accounts. There are however some large business accounts that are typically late, especially if they cannot be shut off (e.g. hospitals, large manufacturing plants with special arrangements). As a result, delinquency among business accounts is much higher, about 80 accounts over 90 days out of 600 accounts.

This whole system is software-driven; all the late notices are automatically generated. This system has eliminated much manual work as well as being more reliable. Previously, an employee would select delinquent notices, which introduced subjectivity. Now there is just one employee involved.

A major benefit from this approach is that the vast majority of bills are all paid before the next reading cycle, avoiding any confusion regarding delinquent payments that extend over multiple billing cycles.

Louisville Water Company

Louisville Water Company (LWC) is a public water utility owned by Louisville Metro Government. LWC provides water service on a retail and wholesale basis to over 1 million residents/231,000 residential accounts.

LWC operates two walk-in bill payment facilities. The corporate location has five full time cashiers that assist an average of 1,100 walk-in customers per day with a dollar volume of $300,000/day. The satellite office in Bullitt County accepts bill payments only. Cashiers are cross-trained with CSR's in the call center.

Over 65 % of all customers pay by mail with the automatic bank draft gaining in popularity. One reason for increased use of this service is the added feature of the customer having the ability to select which day of the month the payment can be debited from their bank account to cover their bill. Customers with final bills and late bills have the option of paying by credit card over the phone.

LWC's low unit cost per walk in transaction is due to the high volume of simple over the counter transactions, mainly payments, as well as well-trained staff. Being a high volume operation also helps keep the waiting time low, compared to other utilities that have only one person who serve customers in person and thus have a harder time controlling the queue.

Cross training with the call center helps match supply of employees with demand fluctuations, and also is a reason that this function cannot be outsourced. Each cashier is trained to handle all potential transactions and since the work volumes are high they get a lot of practice in them. Cashiers all start in the call center so they know the billing system well, and they also work closely with the accounting group.

PPL Electric Utility

PPL Electric Utilities provides electricity delivery to 130,000 accounts in eastern and central Pennsylvania and markets wholesale energy in the US and Canada. PPL believes that its staffing levels in credit and collections are typical of North East electric utilities, but it has good collections performance. Many processes are automated. Staff in this area are also very experienced.

Deposits are required from residential customers who do not have a good credit record, but not others. CSR's have a direct on-line link to credit rating agencies, and PPL was the first in Pennsylvania to have this capability. Deposits are also required from customers with a poor payment record, as 40% of uncollectibles are unpaid final payments when customers move. For this reason PPL is promoting direct debit.

PPL also has a customer assistance program to help customers who are having problems paying their power bills. The utility's rate structure includes an approved component to cover such assistance to customers certified as low income, so long as those customers keep paying an agreed amount every month. PPL is able to write off a percentage of the balance owed by such customers.

PPL's own research confirms that its credit control for commercial customers is very good by electric industry standards. The utility has specialist representatives that focus on not allowing commercial customers to become too overdue in their payments. Given the high proportion of revenue from commercial customers, this makes PPL's overall credit performance good. Specialists use alert services from Standard and Poor's and Dun and Bradstreet to become aware of possible collections problems in the making, moving to collect deposits from customers who are giving signs of financial distress. PPL also has found ways such as turning off power to non-essential facilities, e.g. administration buildings, to collect from "essential services" customers like hospitals.

Orange & Rockland Utilities (O&R)

Orange & Rockland Utilities is an electric and gas utility headquartered in Pearl River, NY. O&R and its two utility subsidiaries, Rockland Electric Company and Pike County Light & Power Co., serve a population of roughly 700,000 in seven counties in New York, northern New Jersey and northeastern Pennsylvania.

O&R has implemented a "continuous transfer program" within its Customer Information Management System (CIMS) to match uncollected balances with new customers.

When a customer closes out an account, there is a record of it. If that same customer then opens another active account with O&R at a later date, there is a record of that transaction as well. The continuous transfer program makes the match that provides a link between the finaled or written-off account and the active account.

(continued)

Orange & Rockland Utilities (O&R) (continued)

Prior to the system's creation, there was no automated method of matching active customers to previously closed accounts. The company lost money when customers didn't pay the final amount and then closed their accounts. When some of the same customers opened new accounts with O&R at a later date, there was no seamless way to establish the link between the accounts. The continuous transfer program creates that link automatically.

Now, this automated process is a weekly step. It matches the old customers' finaled or written-off balances with the same customer name, social security number or driver's license number. In addition, the company is using an outside resource called Accurint that assigns a unique identification number to all individuals. These identification numbers have been assigned to all customer records and offer another link to finaled and written-off accounts. The continuous transfer program can also match on that identification number, regardless of the actual names on the account.

If a match is found and there is an unpaid balance on the account, the customer receives a letter advising that balances, credits or debits from a previous account will be transferred to the next bill for the active account. Another important feature of the continuous transfer program is that even after the transfer is made, any subsequent payments or adjustments on the finaled or written-off account also are automatically transferred to the active account.

Before O&R began the program, it anticipated that these letters to the customers would generate concern and, therefore, higher call levels into its Call Center. To help prepare for this impact, several steps were taken: Customer Service Representatives received training regarding this new process; in an effort not to overload the Call Center, the written-off accounts were introduced to the continuous transfer program in six-month increments, and a "ready-to-send" packet of information was created to expedite matters with customers claiming that this was not their account.

During the start up phase, 4,445 accounts representing about $1.5 million were initially transferred, and 80 percent of that total was collected during the first ten months of the continuous transfer program. It has been a great success. Since then, more than $5 million has been transferred, with a similar success rate.

In addition, Call Center impacts were negligible and there has been no increase in customer complaints to regulators. The continuous transfer program has added another dimension to helping the company better manage collections. It also has provided O&R with a professional process to work with its customers on these specific billing matters.

The City of Phoenix Water Services Department

The City of Phoenix Water Services Department provides water and wastewater service to approximately 1.4 million people in Phoenix, Arizona.

This utility offers many ways to pay bills, including, all at no cost to customers:

- The option for payment to be deducted automatically each month from their savings or checking account
- Payment by credit card on the utility's Web site
- IVR Credit Card Payment, 24/7, by use of a touch tone phone
- Bank branch payments. Bills can be paid at any of the utility's Bank's branch offices. (But no delinquent bills or bills with final turn off notices can be paid this way.)

Having these options pleases customers, but short-term costs have increased. Future automation of back office payment processing is expected to reduce costs.

Currently the utility has five Utility Transaction terminals accepting payments at four locations. In addition the utility partners with a valley electric utility company (Arizona Public Service) that uses two Utility Transaction terminals that process both water and electric payments. This provides a one-stop shop for customers who can pay both water and electric bills at one time. Both utilities employ their own personnel and equipment at this location. Payments processed through the Utility Transaction Terminals for fiscal year 2003-2004 totaled 63,668. Plans are in process to add additional terminals in 05/06.

Long Beach Water Department

Long Beach Water Department (LBWD) provides water, sewer, reclaimed water and storm drain services to over 461,000 customers in Long Beach, California.

Since error free bills result in fewer customer contacts, several methods have been developed to deliver the most accurate bill possible. Billing parameters are tight in the field (i.e. closeness to typical usage) but loose in the office. This results in checking of questionable reads at the time of regular reading, but few kick outs in the office.

The billing system produces a variety of error messages that results from different analyses of the account information provided for billing. These include comparisons to previous usage, no consumption, new reads less than the previous reads, and reads not matching information in the system. Selected bills are then diverted to be reviewed and corrected prior to delivery to the customer.

Recent changes in the procedures for investigating high bill complaints have reduced the operational costs for managing these functions by 60%-70%. Historically, each customer contact

(continued)

Long Beach Water Department (continued)

for a high bill complaint followed the same pattern: reread and field investigation with some cases resulting in a meter calibration test and administrative hearings. This system was replaced with a database to gather customer information that is transmitted to Billing and Customer Account Management, which does an account analysis. If the account bills within 10% of the previous usage in the past two years, then a letter is sent to the customer telling them the bill in correct. This results in an almost 90% acceptance by the customers eliminating the field reread and investigation functions (almost all high bill complaints are without merit).

Customers are able to use the IVR system and the Internet to pay their bills (8% of bills are currently paid by either method as of March 2004). The system provides real time access to account information as well as accepting payments using credit and debit cards. The work volume so handled was equivalent to 2 FTEs by March 2004. Other payment options include home banking, automatic payment and levelized payment plans. Under the automatic ("Easy Pay") system, a bill is sent 10 days before payment is due, and if the customer does nothing the payment is automatically deducted from the bank account. About 3% of customers currently use this facility. The levelized payment system uses a 12-month rolling average charge, so accounts are never trued up. However, customers appreciate the smoothing effect on payment. Even though no convenience fees are charged for any of these options, the cost to the utility is less than the equivalent manual processes.

Long Beach's cost per bill is low primarily because of the multiple services billed on one bill (water, sewer, gas, solid waste, recycling and city tax). One challenge is to agree amongst the parties what happens in the event of partial payments.

Long Beach's excellent collections performance starts with obtaining a positive ID for all new customers. This usually involves an SSN check to a credit agency while the applicant is on the phone. Response from the agency comes within 3 seconds. Possible responses are match, no match, multi-match (more than one name associated with the SSN) and fraud alert (the SSN was never issued, the person with the SSN is deceased or the SSN has been used fraudulently). Unless there is a match, the applicant must bring a valid picture ID to the business office to obtain service. Co-tenants are all held equally responsible for bills, so when such service is set up the utility attempts to obtain details on all co-tenants. Such verification has also identified delinquent amounts to be paid (to any City Dept.), and has resulted in a 63% decrease in bad debt write-offs since 2000.

Internal collection staff skip trace all closed utility accounts with an outstanding balance greater than 45 days in an attempt to locate the customer and collect the bad debt. Bad debts remain in the billing system for 5 years, and this catches many people such as college students who leave the City over the summer but return later.

(continued)

Long Beach Water Department (continued)

Tools used include:

- Credit reporting agencies
- Department of Motor Vehicles
- Equifax Cue Exchange
- Mosaix Dialer Calling Campaign
- In-house billing systems
- In-house utility SSN match
- Data Quick, Stewart Title and County Assessors
- Secretary of State's Department
- Various web sites

Unsuccessful searches to locate up-to-date information relating to the customer are referred to an outside collection agency and submitted to the credit reporting agencies. The collection agency is on an incentive program; higher collections leads to a higher fee. Successful searches for up-to-date information are contacted by letter and/or telephone to arrange collection of the bad debt. Larger debts and unsuccessful payment arrangements are processed in the small claims court. Long Beach also shares information with other counties to chase debtors if they move within the state. An important collection tool is that if there is more than $50 unpaid, *both* gas and water services are cut off. Another important tool is that utility debts can be deducted from state tax refunds before the refunds are disbursed. Collection judgments are good for 10 years.

Sarasota County Environmental Services

Sarasota County Environmental Services is a public agency of Sarasota County Government, Florida. The Utilities Unit serves approximately 65,000 meter connections and 47,000 wastewater connections.

There are 2 office locations that receive over the counter payments and offer other services such as signing up for new service. Over the counter payment volume has decreased, since the utility receives approximately 90% of payments in electronic form by the following means:

1) Bank lock box used for mail payments. Payments are transmitted electronically to the utility by 2:30pm each day
2) E-box - Electronic home banking payments which are received with the lock box file each day
3) Bank Draft – There are a little over 10,000 customers on bank draft.

With these payment options, as well as credit card payments (Visa, MasterCard, Discover) by phone, there is still enough work for 2 cashiers, who are kept busy handling 40,000 transactions per year. Also, in line with prevailing wage rates, they receive less per hour ($10) than many utilities in other regions, which partly explains the low unit cost. Over the counter service quality is rated at 81% by customers, a good result.

CHAPTER 6
METER MAINTENANCE & METER-RELATED FIELD SERVICES

TYPICAL PRACTICES

This chapter focuses on meter maintenance metrics and practices, with an emphasis on key metrics and practices used by best performing organizations, and practical recommendations to water utilities. The utilities that provided data included in this study had the following service territory makeup (Figure 6.1).

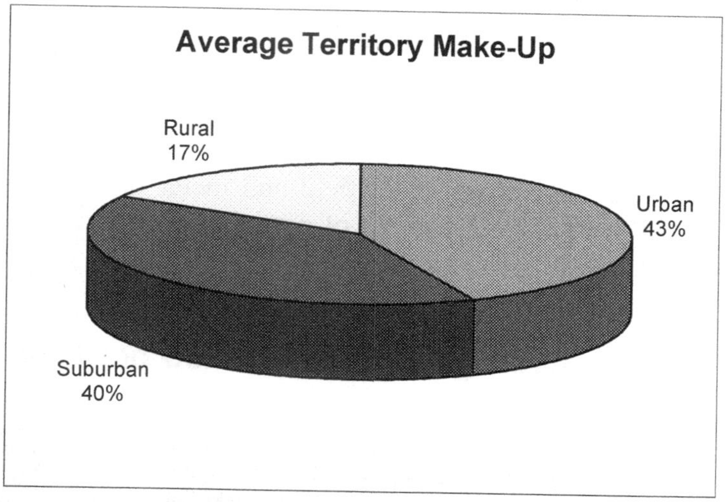

Figure 6.1 Average service territory make-up.

Utilities rely on field forces to visit the customer premise when necessary to install, remove, or maintain the service meter. For the purposes of this study, routine trips to the meter to read the meter (for revenue purposes) is covered in the Meter Reading chapter. This chapter focuses on all other meter-related field service activities required to maintain service delivery. Typical field and meter maintenance activities are listed on the following table (Table 6.1).

Table 6.1 Typical meter maintenance & meter-related field services

Service Orders	Off-Cycle Meter Reads
Turn service on/off	Rereads, check reads
Meter accuracy verification (field test only)	Special reads, pick-up reads
Meter diagnostics	Final reads
Repair meter leaks	Recording voltmeters
Remove/replace/or install meters	Read hydrant meters
Locate meter and curb boxes	

(continued)

Table 6.1 (Continued)

Field Investigations	Other Activities
Billing inquiries and "high bill" investigations	Service branch inspections
Load checks	Various plumbing repairs
Meter resealing	Underground locating
Verify address accuracy	Property service sketches
Back flow investigations	
Water line inspections	
Theft of service investigation	
Water Quality (low/high pressure, dirty water)	
Tap verification (location, size, or active)	

Participants reported an average of 22 field and meter orders worked per day per field employee. As shown in Figure 6.2, 46 percent of participants report that field employees work more than 20 orders per day.

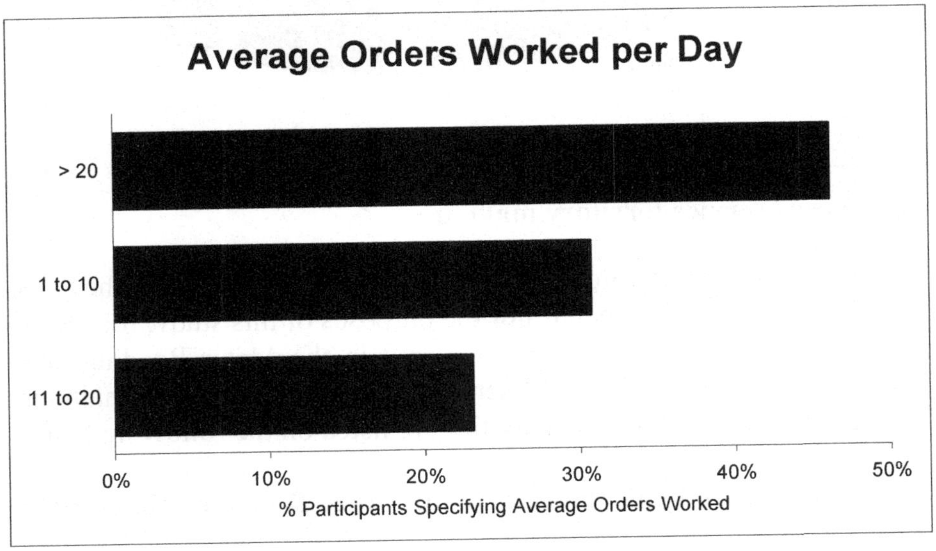

Figure 6.2 Average orders worked per day.

Most participating utilities do not collect payments in the field with field service employees—only 29 percent use their field service employees as collectors (Figure 6.3).

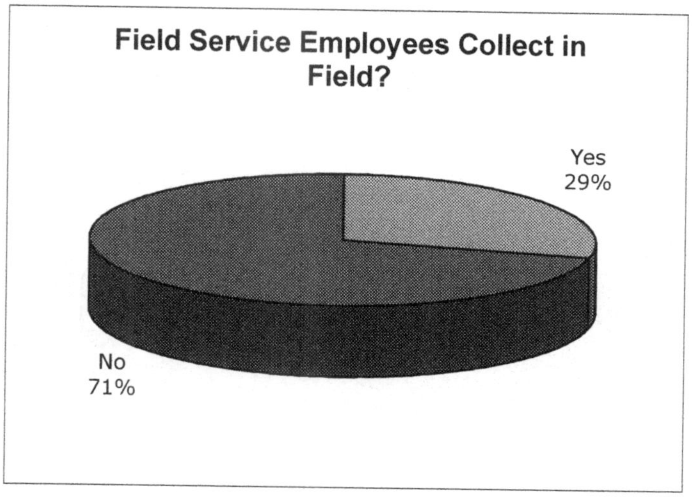

Figure 6.3 Participants whose field service employees collect in field

Almost 30 percent of participants outsource some meter or field service activities to a third-party (Figure 6.4). On average, those companies using a third-party for meter and fieldwork have been outsourcing for 5 years. Participants report outsourcing the following meter and field service activities:

- Meter Replacement
- Meter Installs
- Meter Repair
- Off-cycle Readings
- Special Projects

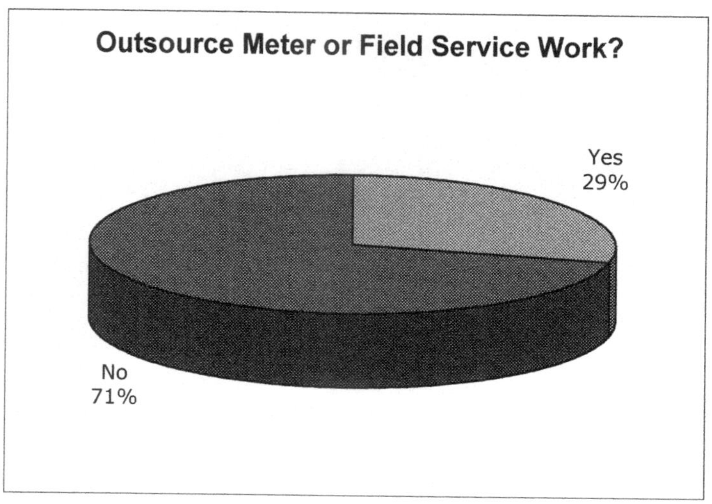

Figure 6.4 Participants that outsource meter or field service work

Half of participants match "move-in" orders to "move-out" orders, whenever possible, for the same address, half do not (Figure 6.5). This technique, which is most easily implemented through automation of the work order system, reduces trips to the meter and operating costs.

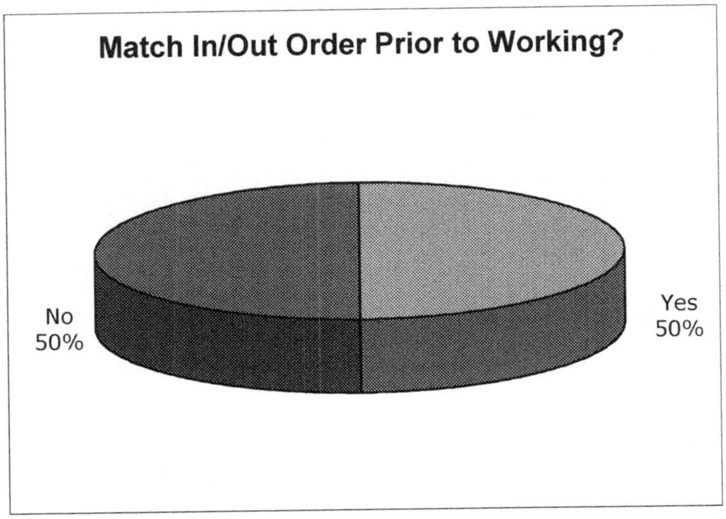

Figure 6.5 Participants that match in/out orders prior to working

Only one-third use schedule-reads as a substitute for final read, pick-up read, or special read orders (Figure 6.6). Those that substitute scheduled reads for special reads report using a 2 to 5 day window to pull these orders. This approach can eliminate a significant percentage of fieldwork, depending upon the seasonality of the business and size of the transient population.

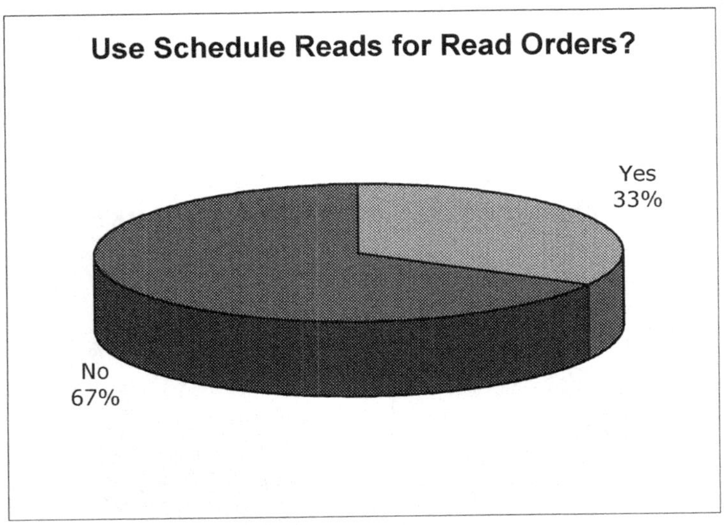

Figure 6.6 Participants that use scheduled final, pick-up or special reads

Some companies are now implementing Automated Meter Reading (AMR) solutions in high-turnover areas, like apartment complexes, to address the large volume of in-out orders at the start or end of each month. Some types of AMR, such as power line communication (PLC) AMR, provide the final and beginning reading without a trip to the meter and expedite the move-in and move-out process for customers.

Most participating utilities offer customers the option of an appointment for customer-requested field service work (80 percent) (Figure 6.7). These include AM/PM appointments, exact time appointments, and appointments within a half-an-hour or hour time window.

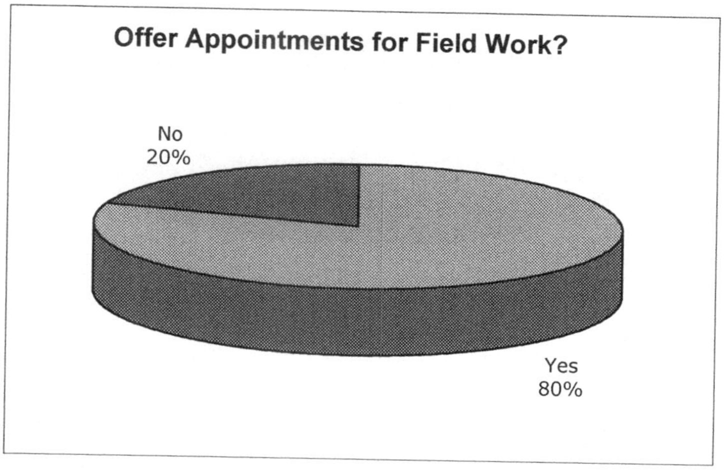

Figure 6.7 Participants that offer appointments for fieldwork

The most popular appointment offered by participating utilities was the "AM/PM" appointment (Figure 6.8). The second most offered appointment type was the "exact time" appointment.

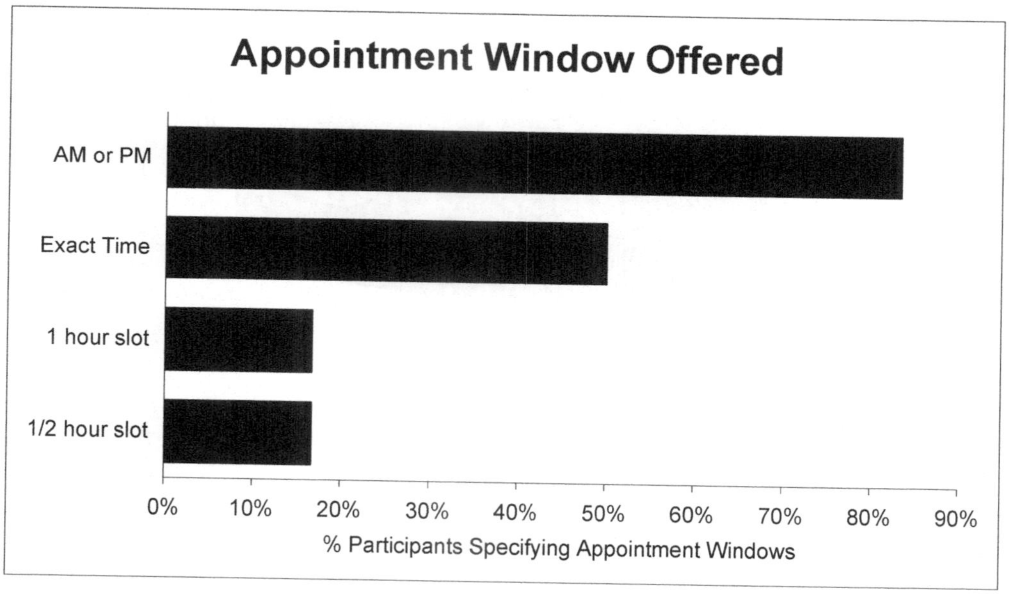

Figure 6.8 Appointment window offered

Of those companies offering appointments, appointments are kept, on average, 95.5 percent of the time. Only 30 percent of participating utilities offer appointments after hours or on weekends (Figure 6.9). None of the companies offering appointments reported any fee charged to customers for normal appointments or after-hours appointments.

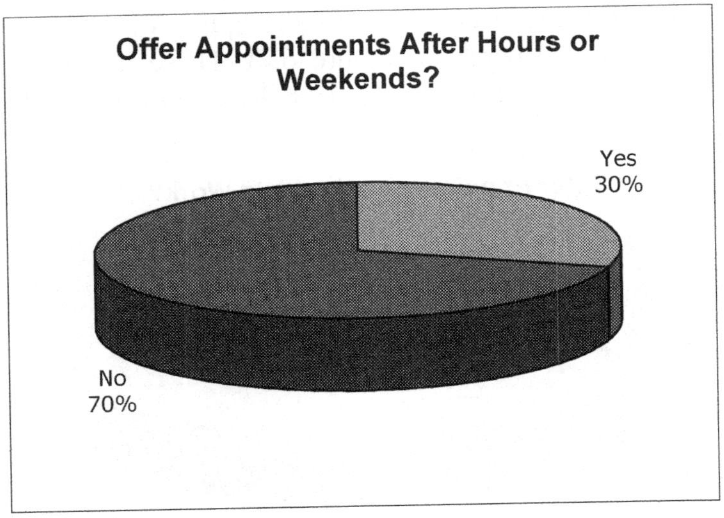

Figure 6.9 Participants that offer appointments after-hours or on weekends

More than half of participants offer some form of "service guarantee" to customers (Figure 6.10). Service guarantees are an important element in establishing service commitment to customers as well as setting service delivery expectations.

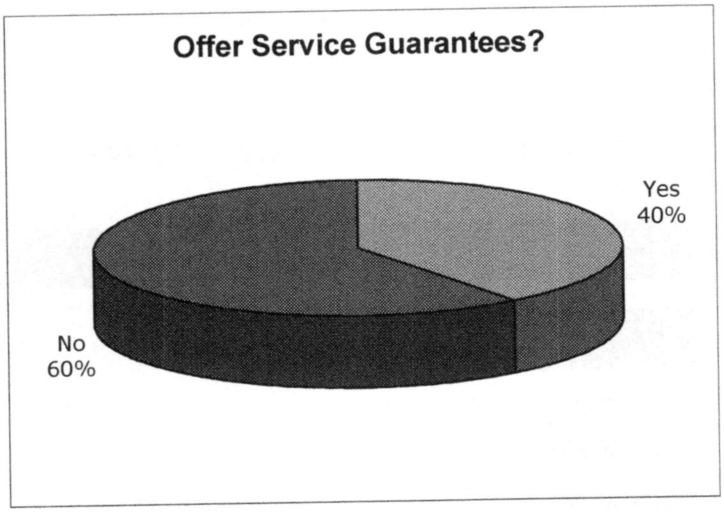

Figure 6.10 Participants that offer service guarantees

Participants reported the following guarantees:

- New service turn-ons within 24 hours.
- Install new service within 7 days.
- All customer-requested work (high bill investigation, final reading, leaking meter) completed on the scheduled day.
- 48-hour turn-around time on rechecks & audit reads.
- Delinquent account shut-off's have service restored within 24 hours after payment.

Participants report an average of 8 days to install a new meter—from customer request to completed installation. Half of participants report an average of 1 to 2 days from customer request to meter installation (Figure 6.11).

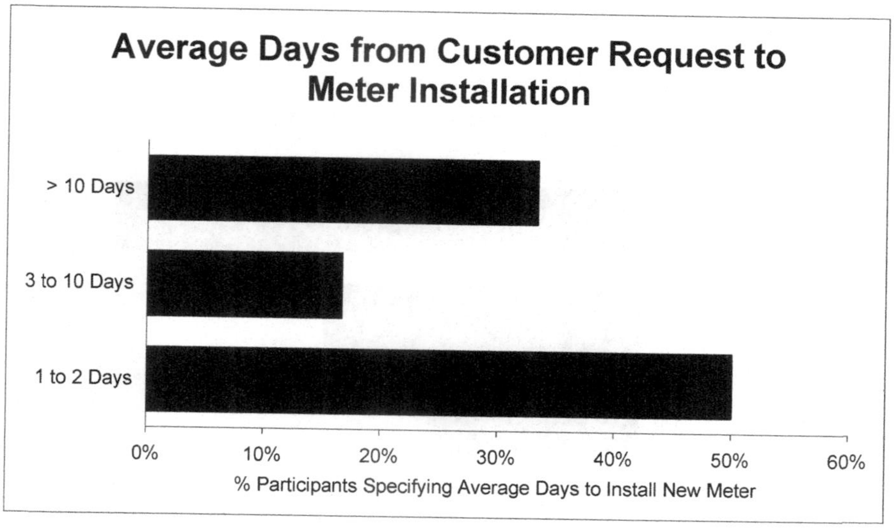

Figure 6.11 Average days from customer request to meter installation

Most participants sample-test new meters for quality control purposes (71 percent). A small percentage (14 percent each) either tests all new meters or relies on the original manufacturer's test certification. Refer to Figure 6.12. The latter approaches maximize the identification of inaccurate meters while minimizing testing costs.

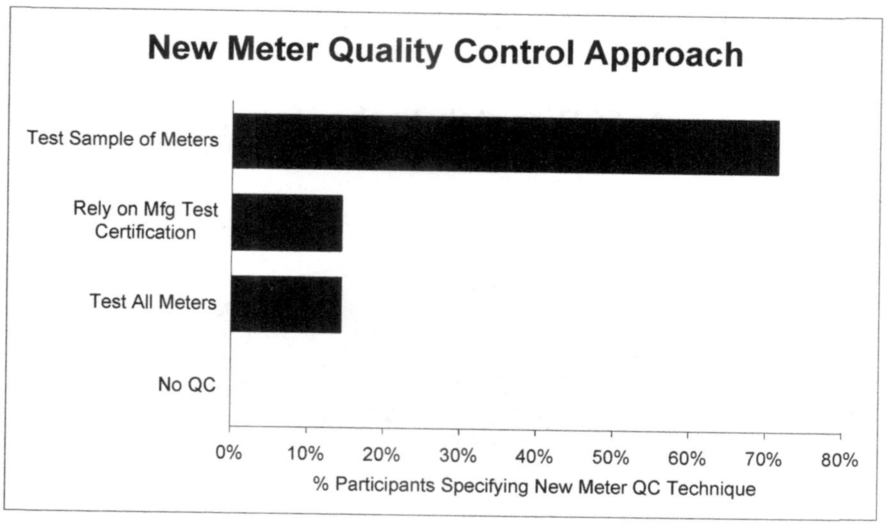

Figure 6.12 New meter quality control approach.

Nearly 90 percent of respondents rely on cell phone technology to communicate with field personnel (Figure 6.13). While 2-radio was the mainstay a few years ago, cell phones technology now dominates utility meter maintenance field communications. One-third of participants use mobile data terminals in the field as their primary field to office communications tool.

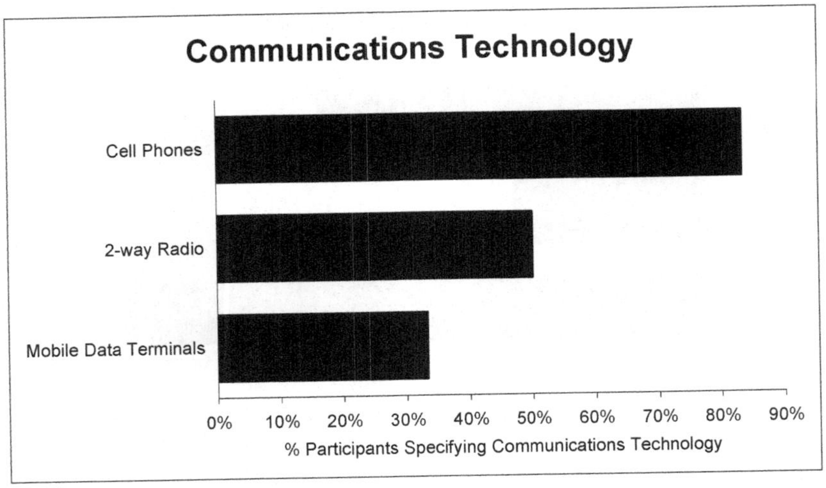

Figure 6.13 Communications technology

Only 17 percent use Auto-Dispatch technology to manage field and meter maintenance work activities (Figure 6.14). Auto-Dispatch technology provides work order management, workload balancing and scheduling, and mobile data communications to facilitate better management of field forces and assets. This technology is more widely used in the electric and gas industry to facilitate emergency response and service delivery.

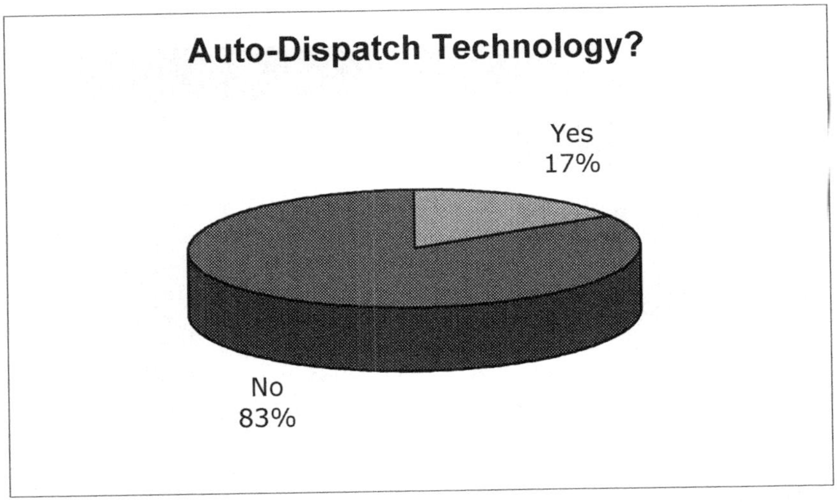

Figure 6.14 Participants using auto-dispatch technology

METRICS AND BEST PRACTICES

The key efficiency and effectiveness metrics were calculated from data supplied by approximately 13 utilities, including 8 water utilities. Performance metrics are provided for two functional areas: Meter Maintenance and Field Services. The figures in the following two tables (Tables 6.2 and 6.3) contrast the best performers (those utilities that had both above average efficiency and effectiveness) with the utility industry average as well as the water industry average.

Meter Maintenance

The key metrics we have chosen for meter maintenance to provide a balance between service and efficiency are:

- Service orders worked per involved FTE (annual)
- % Inaccurate meters (meters tested as inaccurate)
- Unit cost (cost per order worked – operational & maintenance costs only – direct labor, contractor costs, overtime, and non-labor O&M; no capital costs or overheads)

Examining the number of service orders worked per meter maintenance employee is a good measure of individual productivity. This measure can also be used to measure group, team, and overall departmental performance. It is important to keep in mind that some types of service calls take longer, on average, than others to complete. When comparing individual productivity, it is important to look at the types of orders as well as the overall number completed per day, per week, per month, and per year.

To maintain a balanced view of performance, companies were designated as "best performers" if they exhibit above average performance in both cost and service. Once the "best performers" were identified, "best performer average" metrics were calculated, as reflected in Table 6.2 below.

Table 6.2 Meter maintenance benchmarking performance

Meter Maintenance Benchmarking Performance			
All Utilities	**Best Performer Average**	**Utility Industry Average**	**Water Industry Average**
Orders per FTE	5,277	2,842	4,484
Percent Inacccurate Meters	0.07%	3.24%	5.11%
Cost per Order	$ 17.45	$ 46.29	$ 20.17
Utilities Participating	2	7	3

Meter-Related Field Service

The key metrics chosen for meter-related field service to provide a balance between service and efficiency are:

- Service orders worked per involved FTE (annual)
- % Service orders worked on time
- Unit cost (cost per order worked – operational & maintenance costs only – direct labor, contractor costs, overtime, and non-labor O&M; no capital costs or overheads)

One of the best measures of field service effectiveness is the percentage of service orders that are worked when promised. This is fairly easy to track, provided the company maintains certain standards of service (eg connect within 24 hours, disconnect by the next day). It becomes a critical measure when a company chooses to offer a service guarantee program that emphasizes keeping customer commitments. Another similar measure that can be tracked and reported is the number or percentage of appointments that are kept vs. missed. Neither measure is intended to measure individual performance; they are intended to measure the performance of a large group, region, or an entire field service organization.

To maintain a balanced view of performance, companies were designated as "best performers" if they exhibit above average performance in both cost and service. Once the "best performers" were identified, "best performer average" metrics were calculated, as reflected in Table 6.3 below.

Table 6.3 Meter-related field service benchmarking performance

Field Service Benchmarking Performance			
All Utilities	**Best Performer Average**	**Utility Industry Average**	**Water Industry Average**
Orders per FTE	7,887	5,018	5,019
% Orders Worked on Time	95.00%	92.75%	92.57%
Cost per Order	$ 4.91	$ 14.30	$ 14.47
Utilities Participating	3	8	7

Most utilities are focusing on one or more of the following approaches to meter maintenance and meter-related field service improvement:

- Reducing the cost of a trip to the meter through technology and automation:

    - Better work order management;
    - Route optimization;
    - Automated order completion and status update;
    - More sophisticated hand-held and field testing equipment; and
    - Advanced metering technology.

- Eliminating field trips through AMR, prorating cycle reads and combining same-account orders, whenever possible.

- Contracting meter maintenance and installation to reduce overhead, tackle seasonal peaks, or as a strategy to transition to automated meter reading.

Best performing meter maintenance and meter-related field service companies are:

1. *Using Mobile Workforce Management Systems to Improve Efficiency and Effectiveness.* Today's mobile workforce management systems include automatic tracking of service order performance. Management defines the rules and service levels and the system manages the work accordingly, recommending when the work should be assigned and who should be assigned the work. These systems can track commitments and appointments kept or missed as well as cost per order and field service employee productivity. Most of these systems also include automatic time-keeping features so that employees do not have to keep up with paper timesheets.

2. *Implementing Clear and Concise Measures of Performance* – The "best performers" identified in this study were deliberate in their measurement of employee, group, and departmental performance—cost, service, and productivity. Best performers reported providing employees with a clear idea of job expectations and performance. Work management systems were used to assign and track work status as well as employee productivity.

3. *Prioritizing Customer-Request Orders over Internally Generated Orders* – The "best performers" identified in this study use automation to prioritize work orders. Emergency and customer-requested work orders have the highest priority to ensure safety and service response.

4. *Deploying Cellular Phones to Improve Communication Between Field and Office.* Equipping field and meter service personnel with cellular phones provides an efficient and effective means of communication to ensure better service response. Field employees no longer have to compete for bandwidth on a crowded company radio network. Employees can also contact customers directly to avoid "can't get ins" and missed appointments. Employee to supervisor communication is also enhanced.

5. *Matching Read-In and Read-Out Orders and Other Multiple Trip Orders* – Best performers use automation to match multiple field orders for the same meter (within a certain time-frame) to eliminate repeat trips to the meter. Matching read-in and read-out orders for the same meter can significantly reduce costs and improve effectiveness.

6. *Letting Field Employees Start Work Day From Home.* Mobile data terminals and other field order management systems make it possible to equip employees so they can start the day from home, eliminating unnecessary travel into the work location prior to starting the first job. This makes sense when the employee lives as close or closer to the territory to be worked than the reporting work location. In addition to reducing travel time, this also provides a boost to employee morale. Similarly, eliminating the need to report into the work location at the end of the day can improve productivity.

7. *Cross-training field employees so they can perform all meter related services as well as leak repairs, leak surveys and pipe locating.* Best performers improve the flexibility of their field service staff through cross training and development. This approach creates a multi-skilled employee that can usually resolve multiple service issues on the first trip to the meter.

8. *Emphasizing Safety and Enhancing Safety Training.* High performing companies continually emphasize the importance of safety and safety training. A high-profile safety message and safety culture keeps safety in minds of employees. Including safety metrics in a bonus or incentive program also emphasizes its importance to employees.

CASE STUDIES

City of Phoenix Water Services Department

The City of Phoenix Water Services Department provides water and wastewater service to approximately 1.4 million people in Phoenix, Arizona.

The City hires contractors to repair large meters. There is a strong focus on large meter accuracy. Small meters are replaced not repaired.

Implementation of a backflow policy requiring all contractors/builders to install a back flow device in line with the fire hydrant meter has prevented contamination of the public water system.

Field employees are cross-trained so they can work in various job duties, which provide a flexible work force. These "Water Services Specialists" are combined water inspectors and customer service representatives.

Field and office employees use Nextel two-way radiophones. This provides staff with the ability to communicate directly, which has decreased the number of radio interruptions with the dispatcher. Employees in the field are provided cellular phone access and capability to call or return calls and set up appointments with customer while in the field. Compared to the previous practice of attempting to call customers in the morning and often not being able to, using cell phones as needed has greatly improved the utility's ability to serve its customers.

Future plans include an Automated Field Work Order System (AFWOS), which will facilitate bi-directional delivery of essential data and information between the field service employees and call center specialist through a mobile system. The system will be connected to or integrated with current systems in Customer Service Division and other systems in the Water Services Department. Service to the customer will be enhanced by improving operational efficiency in the office, the field and other Water Divisions by providing real time information regarding status of a work order in the field and the ability to transfer information to other systems in other division of the Water Services Department. This project is scheduled to begin in 2005/2006.

Yarra Valley Water

Yarra Valley Water (YVW) is a state government-owned Australian water/wastewater utility serving 680,000 accounts in Melbourne, Australia's second largest city.

YVW's meter services have been contracted out for 10 years. Initially this involved many separate contracts involved in the meter life cycle. i.e. supply, installation, testing, replacement, repair. However, these contracts have been aligned and consolidated into a "meter fleet" management contract. Under this contract the contractor runs the day-to-day operations in the most cost effective way based on the utility's policies and strategic direction and a schedule of approved rates.

The transition to contractors from unionized workforce was difficult. A 6-month strike occurred, but managers went into the field with contract plumbers, and completed the necessary tasks.

As the contractor provides similar services to other utilities, YVW benefits from the contractor's economy of scale that it can generate through purchasing agreements with manufacturers.

YVW only maintains large meters. Smaller meters are just changed out as needed. If a customer requests the meter to be checked, its accuracy is now tested at the customer's premises, and 99.9% of the time meters are found to be working properly.

For new connections, customers (plumbers, developers) complete an application through a franchised plumbing store (no need to come to YVW's office). The application is processed on line, fees are paid and a work order is electronically logged with the contractor to install the meter.

On completion, the work order is closed (with new meter data) and the file is electronically interfaced with the billing system so that all new meters are loaded within 24 hours of the meter being installed in the field. Closure of the work order also automatically generates an invoice based on the schedule of rates that is sent to the contract manager for electronic authorization.

Other meter-related field service activities now form part of the "fixed price" meter reading contract and are undertaken by the meter reading contractor. Consequently, the contractor has a greater interest in data accuracy, since rechecks and audits add unnecessary cost to the contractor.

Louisville Water Company

Louisville Water Company (LWC) is a public water utility owned by Louisville Metro Government. LWC provides water service on a retail and wholesale basis to over 1 million residents/231,000 residential accounts.

LWC has field service customer service representatives on duty between 7am and 10pm, Monday through Friday, and 7am to 3pm on Saturday. These sixteen CSRs complete an average of 35 service orders per eight-hour shift. For new customers and turn-on requests, there is a same day response. Most inspections are scheduled for the CSR to meet with the customer. In addition to the CSR personnel, the department has 2 leak surveyors and 2 meter exchangers.

Senior staff work as CSRs, and are empowered to make decisions on workflow. It is a highly paid group but productivity is also high, leading to a low average cost per work order. Expectations are high re customer skills as well as technical skills, and the group has a high degree of employee satisfaction.

This area has recently implemented a mobile dispatch system that automatically assigns service orders. The system estimates travel time as well. After completing the service order, the CSR enters all of the information pertaining to the job into the system so it is entered in real time. This eliminates the need for office personnel to key in information and reduces paperwork substantially.

CHAPTER 7
IMPROVING CUSTOMER SATISFACTION

This chapter covers two topics:

- The development and testing of a customer satisfaction survey that any water utility can use to measure its customer satisfaction, track it over time, and compare itself to other water utilities.
- A review of recent research on customer expectations and drivers of customer satisfaction.

CUSTOMER SATISFACTION SURVEY

This survey was developed as described in Chapter 2 Approach, and can be found at www.wisebenchmarks.org. Results for the five water utilities that used the survey were encouraging, with a 10-20% return rate from their customers surveyed.

Although this sample size is too small to conduct analysis, the metrics covered are described below. The survey is available as a self-service tool for water utilities on the project web site. Utilities that enter their survey results obtain instantaneous feedback on how their customer satisfaction compares to averages (which include their own data) in the database, per the metrics shown below.

Average results from the five utilities that administered the survey appear below. These results are not shown because they have any statistical validity (the sample size is too small), but to describe the service aspects with which any water utility can compare itself to water industry averages.

- Overall Customer Satisfaction Rating: 84.9%
- Call Center Courtesy and Professionalism: Overall Rating 82.8%
- First Call Resolution Rate: 59.3%
- Interactive Voice Response Ease of Use: 74.3%
- Interactive Voice Response Time to Agent: 75.5%
- Field Service Courtesy and Professionalism: 86.8%
- Satisfaction with Field Service Workmanship: 86.2%
- Satisfaction with Time to Provide Field Service: 84%
- Overall Satisfaction with Field Service: 87.1%
- Billing Accuracy: 87.2%
- Ease of Understanding Bill: 86.9%

- Over-Counter Service Quality: 82.9%
- Drive-up Service Quality 88.1%
- General Satisfaction:
  - Easy to Do Business With: 86.3%
  - Treats Me Like A Valued Customer: 84.9%
  - Attitude Toward Customer Service: 84.9%
  - Knowledge of Customer Service Representatives: 84.9%
  - Ease of Locating the Right Person: 82.4%
  - Accuracy of Information Provided: 85.3%
  - Promptness & Efficiency in Resolving Issues: 82.9%
  - Customer Service Experience, as Compared to Other Companies: 86%

RECENT RESEARCH INTO CUSTOMER EXPECTATIONS AND SATISFACTION WITH WATER UTILITIES' CUSTOMER SERVICE

Recent AwwaRF research[1] shows that water utility customers expect:

- A focus on customers' needs, eg being listened to
- Courtesy and respect, eg not calling customers by their first name
- Knowledgeable representatives, who also do not pretend to know something they do not
- Good quality water (but taste is not equated to safety, that is, even if people do not like the taste they do not automatically conclude it is unsafe to drink)
- Complete restoration of property when any on site work is done
- To be able to speak with a live person rather than being forced to use an automated system.

AwwaRF research also revealed that utility managers tend to overestimate their consumers' level of satisfaction. For three basic areas of satisfaction (tap water quality, overall customer service, and complaint resolution), managers overestimated satisfaction by 14% to 20%.[2]

---

[1] Source: Developing Customer Service Targets through Assessing Customer Perspectives, AwwaRF, 2003, p xix

[2] Source: Consumer Perceptions of Tap Water, Bottled Water, and Filtration Devices, AwwaRF, 2003, p 89

How Does Customer Satisfaction With Water Utilities Compare With Other Organizations?

Customer satisfaction in the US today is the highest since the second quarter of 1995, according to a report released in Feb 2004 by the American Customer Satisfaction Index (ACSI). The ACSI, produced quarterly, measures customer satisfaction with a wide range of goods and services consumed by Americans. Ratings on the Index are produced through a scientific model in which overall satisfaction scores reflect performance in areas that customers identify as being most important. Overall satisfaction ticked up .3 percent to 74%, continuing a strong two-year upward trend in how people evaluate the quality of goods and services on which they spend their money.

And in December 2003, the ACSI's annual Federal Government special report showed that Americans are more satisfied with government than is commonly thought. Federal agencies registered an average score of 70.9 on the ACSI's 100-point scale. The report found though that agency performance varies, with timeliness and ease of service needing the most work. Courtesy and professionalism rate highly, in the 80-82 range. Expectations are low, an average of 68 as compared to 78 in the private sector. But the real sore point for citizens is timeliness of service (70) and ease of doing business (74). The timeliness score plunged downward, off by three points since last year.

Satisfaction with public drinking water quality is not measured by either of the major customer satisfaction indices: the ACSI and JD Power and Associates. But the ACSI finding on timeliness agrees with recent AwwaRF research. A 2003 report on customer perspectives found that customers rated timeliness, a reasonable appointment window, and knowledgeable representatives as very important service attributes for water utilities. The 2003 AwwaRF research found that consumers gave water utilities an overall score of 73 out of 100 for service, putting water utilities roughly on a par with the average private sector organization and slightly higher than the average federal agency. However, the taste of tap water received criticism. Interestingly for the water sector, soft drink beverage leaders have much higher than average customer satisfaction at 83-89%, according to the ACSI. Detailed attributes from this AwwaRF research on water utility customer service perspectives[3] are shown in Table 7.1 on the following page.

AwwaRF research[4] also finds that customers who say they are "very informed" on water quality issues are more likely to be satisfied with their utility's performance. Informed customers, in turn, are less likely to report errors on their bill, problems with taste and odor, or safety concerns with their tap water. This finding suggests that communicating with consumers to educate them may increase customer satisfaction, as well as countering media inaccuracies.

---

[3] Source: Developing Customer Service Targets through Assessing Customer Perspectives, AwwaRF, 2003, pxx
[4] Customer Attitudes, Behaviors and the Impact of Communication Efforts, AwwaRF 2004, p xvi

Table 7.1 Water utility customer service perspectives.

Service Attribute	Score (out of 100)
Billing schedule	82
Authority of reps: customer service	82
Arrival etiquette: field service	81
Professionalism: customer service	80
Professionalism: field service	80
Resolve billing problems	78
Line wait: customer service	75
Phone hold time: customer service	74
Follow-up: field service	74
Frequency utility communicates	70
Hours of operation: customer service	70
Hours of operation: field service	70
Response time for field service	68
Appointment window: field service	68
Water quality info schedule	67
Automated Phone: customer service	64
Internet information	63
Average	73

J.D. Power and Associates' 2003 Gas Utility Residential Customer Satisfaction Study is an annual study measuring customer satisfaction performance among 55 of the largest local gas distribution companies in the country. The overall customer satisfaction measurement is based on five factors (in order of importance): company image; price and value; billing and payment; customer service; and field service.

The study found that customer satisfaction is highest among those who believe their gas utility performs well in providing pricing and payment options. And while 75% of consumers pay their gas utility bill through the mail or by visiting their utility, 90 % of customers would prefer to pay through an alternative method, such as the Internet, payment agency or automatic deduction from their bank account. When customers pay their bills over the Internet, they tend to rate their utility's Web site higher. The study also finds that consumers are open to managing bill amounts through energy efficiency and rebate programs, which may keep potential rate increases from impacting their utility bills. In this regard, real time AMR systems offer the potential for customers to manage usage and enhanced satisfaction with their utility.

CUSTOMER SATISFACTION STRATEGIES IMPLIED BY THE RESEARCH

Organize Processes with Customers in Mind

A 2001 American Customer Satisfaction Index (ACSI) report implied that water utilities' customer satisfaction can increase by emulating practices of leading private sector organizations: it found that the IRS had improved its satisfaction rating by adopting some of the private sector's techniques, such as actively listening to their customers, then changing processes, reallocating resources, and focusing technology to enhance the overall tax-filing experience.

Part-time agents sometimes help provide coverage during lunches and breaks. Energy utilities such as Public Service New Mexico, and some water utilities such as Spartanburg Water System have CSRs handle walk-in traffic and paperwork as work volumes ebb and flow during the day.

Provide Multiple Payment Options

If similar conclusions can be inferred from JD Power's research on gas utilities, improving payment options and the prospect of managing water usage through real time feedback may offer significant boosts to customer satisfaction. There is no reason to believe that water utility customers do not value the customer service benefits provided by expanding payment options.

Use the Internet and IT to Improve Service and Reduce Cost

Leading energy utilities have made technologies such as Web self-service, interactive voice response (IVR) and automatic meter reading (AMR) a focus of improvement efforts. Energy utility customers now expect to be able to access information regarding accounts, view and pay bills, prepare a home energy audit, and obtain information on energy-saving appliances or home modifications. Energy utilities have also used this expansion of payment options and self-service information as an opportunity to close business offices.

Internet billing and flexible payment options are increasingly common within the water industry. However, use of IVR is slower than adoption of the Internet for billing and payment. Some water utilities are using IVR to provide information such as bill and payment information and directions and office hours, and many more sophisticated features such as the ability to extend payment due date, enter a meter reading, and make a change to property ownership details. Involving customers in IVR system design has been shown to improve their satisfaction with the system, as does measuring IVR performance regularly.

Improve Customer Contact Handling

Customers are happiest when their calls are resolved on the first call, even if the call is longer. Leaders in customer service provide comprehensive information to customer service representatives (CSRs), including instantaneous meter reading where the type of AMR system allows it (eg powerline carrier), as well as access to other systems such as work management systems.

The average speed of answer is also important, and the acceptable time seems to be coming down. Private sector call centers aim for very low answer speed, in just a few seconds, while leading utilities aim to answer calls in 20 seconds or less.

Facilitating a high level of service is a trend towards centralizing call centers (including virtual centralization). For example, award winner PPL Electric Utilities has one virtual network connects three physical call centers. The three centers share one common Automated Call Director (ACD) and one IVR. CSRs are responsible for customer care, billing, collection, new/upgraded service, outage reporting and dispatch, and also perform back office work.

Call quality monitoring, using that monitoring as feedback to coach CSRs, and having dedicated call QA personnel are mechanisms used by service leaders.

A perception of employee indifference is one the main reasons customers leave a supplier and go to a competitor. Therefore, all employees who may come into contact with customers must be attentive to their needs. In this regard, many more employees than those who work in the core customer service functions need to be customer-aware.

Customer Service Employee Selection, Training and Support

Actions taken to improve the customer experience on the phone include better selection and training of CSRs, close monitoring of call satisfaction, feedback/rewards to CSRs, allowing CSRs to work at home, and otherwise having an emphasis on making CSRs feel valuable.

Service leaders place a great deal of emphasis on selecting the right service personnel, for example through role-playing.

Ongoing performance measurement at the individual level is a practice common to customer service leaders. Best performers set definite standards, providing training and coaching and linking salary, bonuses and other rewards to individual and group service performance and efficiency. Individuals in such leading companies receive frequent feedback on their performance, often daily.

Effective first line supervision is a common theme for service leaders. Some research indicates that a poor relationship with the supervisor is the main reason employees leave. Also, under the right circumstances, leaders are allowing some CSRs to work from home. This practice can reduce costs, open up opportunities for people who want to work part time, and improve employee satisfaction generally.

Adopt Automatic Metering

There is a trend to AMR but currently less than 10% of water utilities have it. Utilities are realizing both tangible and intangible benefits from this technology. Depending on the technology chosen, benefits can include reduction of estimated reads (which reduces the call center workload considerably), ability for high bill complaints to be resolved by the CSR while the customer is on the phone, the ability to read meters as frequently as desired, improved accuracy of large meter reads and elimination of manual checks, automatic theft of service detection, and leak detection on customers' premises (some of these benefits are only available with a fixed network system, not drive-by).

Provide Special Services for Business Customers

Business customers have even less patience with waiting time than residential customers, and have specialized needs. Customer service leaders strive to answer calls from business customers with almost no waiting time. Leading utilities are providing business customers with dedicated CSRs, special toll-free numbers to dial in, bypassing the IVR and connecting directly to business specialists that function as key account representatives. Business customers are also more attuned to voice mail than waiting in a queue, so providing that option is valuable to them as long as they know they are dealing with specialists and especially if they know the person to whom they are leaving the message.

Other special services for business customers include Web-based methods for builders and contractors to submit requests for service to new homes, to obtain an easy way to check on the status of service requests, to update common job status information and to provide information about issues or concerns related to the installation of services.

CHAPTER 8
CASE STUDIES FROM CUSTOMER SERVICE LEADERS

This chapter provides case studies on 9 excellent customer service organizations. These organizations all demonstrate high service and efficiency via the metrics and practices surveys conducted during this project, and/or by reputation (primarily JD Power or American Customer Satisfaction Index leaders).

There are a number of common themes from these case studies, including:

- Selecting the right customer service representatives (almost all cases)
- Training of customer service representatives (Continental Airlines, State Farm Insurance, Dell, Louisville Gas and Electric, PPL Electric Utilities)
- Actively seeking feedback from customers (State Farm Insurance, Dell, Yarra Valley Water)
- Using the latest customer service technologies such as IVR and web-based self service (Continental Airlines, Dell, Fedex, PPL Electric Utilities, Washington Suburban Sanitary Commission, Yarra Valley Water)
- Focusing on employee satisfaction, development and career opportunities (Continental Airlines, State Farm Insurance, Dell, Fedex, PPL Electric Utilities)
- Call quality monitoring (State Farm Insurance, Dell, Louisville Gas and Electric, Power New Mexico)
- Having a separate business customer call center (Dell, Louisville Gas and Electric, PPL Electric Utilities)
- Focusing on first call resolution even at the expense of productivity measures such as the number of calls handled per representative per day (Dell, Power New Mexico)
- Having incentive programs for customer service representatives (Dell, Power New Mexico)
- Having as many payment options for customers as possible (Louisville Gas and Electric, Washington Suburban Sanitary Commission, Yarra Valley Water)
- Automatic meter reading (PPL Electric Utilities, Washington Suburban Sanitary Commission)
- Using a mix of in-house employees and contractors, pooling different groups of employees and/or using spare capacity to provide services for others (Continental Airlines, State Farm Insurance, Louisville Gas and Electric, Washington Suburban Sanitary Commission, Power New Mexico)

- Having programs to assist customers who have difficulty paying their water bills (Washington Suburban Sanitary Commission, Yarra Valley Water)

Other notable practices include having a mix of performance measures for customer service representatives (Continental Airlines); after hours response (State Farm); having a strong focus on credit and collections performance (Louisville Gas and Electric); using mobile dispatch for field workers (Louisville Gas and Electric); using meter route optimization software (Washington Suburban Sanitary Commission); having all field workers under the customer care organization (Washington Suburban Sanitary Commission); and providing customer service guarantees and rebates for poor service (Yarra Valley Water).

CONTINENTAL AIRLINES

In January 1995 Continental Airlines' CEO Gordon Bethune unveiled his four-point "Go Forward Plan" designed to improve operational performance and the working environment for employees. By March, Continental had achieved the top-spot in domestic on-time performance for the first time since the Department of Transportation began tracking performance in 1987. By August Continental placed first in domestic baggage handling. At the close of 1995, Continental announced the largest annual profit in its 61-year history. In addition, the airline ranked first in on-time performance, baggage handling, and second fewest in customer complaints. Continental's Go Forward Plan had launched one of the greatest corporate turnarounds of the 20th century.

Since 1995, Continental continues to receive the highest recognition for superior service, baggage handling, and on-time performance. Consistently ranking at the top in terms of customer satisfaction, Continental Airlines believes the consumer experience begins the moment a call is placed to one of its four reservation centers. An efficient, responsive staff supported by state-of-the-art technology and a company focus on employee morale helps to ensure that the Continental customer's experience with the airline is successful and positive.

"Continental has maintained one of the top reputations in our industry because we are focused on what customers want — clean, safe and reliable service," said Gordon Bethune, chairman and chief executive officer of Continental Airlines. "Our dedicated employees, who deliver quality service every day, have earned respect from our customers, investors and peers."

New Technologies Pave the Path

Continental's reservation centers, including two in Houston, one in Tampa, and one in Salt Lake City, are fully networked into one virtual center. Approximately 3,700 full-time equivalent non-union employees—translating to roughly 5,000 employees—staff the centers, handling 65 million inbound calls in 2001. According to Larry Goodwin, Continental's vice president of reservations, "For the last three or four years, we have been using the Cisco ICM system to route calls to the next available agent and managing the four centers as one big center. That has been a big boost from a productivity standpoint."

Continental agents answer each call directly then transfer as needed, to special desks or to the Integrated Voice Response (IVR) unit. In fact, Continental does not and has never used an upfront routing menu. When the Reservation Centers had the technology that could provide the prompt, management made a conscious decision not to put the prompts in place. "This was for the customer's best interest, to get them where they want to go in the quickest amount of time," comments Goodwin.

When Continental falls below its service level—usually due to a weather-related situation—an agent still answers the phone directly. "Customers really like this," comments Goodwin. "It's so very different from experiences with other call centers.

Continental offers bi-lingual, speech-activated, self-service options through its IVR system. Callers can obtain flight information, check schedules, and access One Pass frequent flyer accounts without speaking to an agent. The IVR also offers self-service booking options for employees needing to travel within the Continental system. Of the 65 million inbound calls in 2001, approximately 8 million were handled completely through the IVR, with the flight status application handling the majority of the calls.

A few years ago, Continental overhauled its reservation center technology, replacing workstations and adding a new front-end to the reservation system, Computer-Telephony Integration (CTI), and speech recognition. The new technology has made it easier for agents to respond to callers.

Recently, Continental began using predictive dialers to alert customers of schedule changes. "The dialer has been a lifesaver for us," Goodwin comments. "Especially after September 11th, in which there were a number of schedule changes in a very short time period." Continental used its predictive dialer to call customers whose schedules changed, relaying a message for customers to call a special toll-free number. Other new technology recently implemented includes workforce management and scheduling software, Witness Systems' eCoach call monitoring solution, and an e-mail management package.

Continental also has a new, internally developed, front-end for the reservation center, designed to provide agents with a user-friendly interface in a Windows-based environment. Tagged EZR, the new program includes built-in wizards to assist agents in performing the more exotic transactions that occur from time-to-time. EZR's wizards eliminate calls to support desks in handling these infrequent transactions. Continental's Reservation Centers now respond to email inquiries through its Electronic support desk. Approximately 100 customer service agents process e-mail, as well as provide voice support to customers using the web site. Continental has also been experimenting with a click-to-talk technology, running about a 35-to-40 percent conversion rate on calls that originated through the click-to-talk. "It's a homerun for us right now," comments Goodwin.

Encouraging Higher Productivity and Better Quality

For the past three years, Continental achieved distinction as one of the 100 best places to work, according to Fortune magazine. Last year, Continental was the only airline on this list. During this period, the airline's out-the-door turnover was less than five percent, with a current four percent rate. Internal turnover, encouraged by Continental management, averages approximately 15 percent.

Several years ago, Continental initiated employee cross-training through a process tagged "generalization." Reservations agents are trained to handle many different types of calls. This, in combination with the intelligent routing technology, maximizes employee productivity. "Four or five years ago, customers calling with a frequent flyer program question would call a separate group of agents," says Goodwin. "Now, the same agent can help you with reservations as well as frequent flyer questions. This not only helped productivity, but also made the customer much happier, creating more of a one-stop-shopping experience." While reservation agents are generalists, there is still some specialization within the center. Special-function desks support international sales, travel agents, and 13 different foreign languages.

Selecting the Right Agents

Several years ago, Continental developed and implemented an agent profile to aid in the selection of representatives. Currently, the Company seeks to automate the initial process and is testing computer-based applicant screening in its Tampa office. Eventually, Continental hopes to move the process to the Internet so that the initial application can be completed online. Presently, applicants come to the Tampa office to use the computer-based screening tool that enables prospective candidates to either pass or fail the first step in the screening process. Those who pass then proceed to team leader or management team interviews. Each Center hires and trains its own agents.

Agents receive training in both telephone sales and customer service through Continental's in-house training program. At this time, the Training Curriculum and Content are catalogued on the Internet, while Continental continues to develop the Internet delivery system for the training. Continental uses blended learning to teach new reservation agents how to handle typical customer calls as well as navigate through the reservation system. Through a mix of on-line courses and instructor-led training, agents can learn and practice a skill until mastered. The on-line instruction has been so successful that Continental has been able to reduce the amount of time spent in the classroom.

Continental focuses on four performance management measures for each agent—calls per hour, bookings or sales per hour, percent of e-ticket sales, and call quality as measured by Witness Systems eCoach. Agents can be rated above, at, or below target. Depending upon the rating, employees may be monitored more or less the following month. This helps focus frontline supervision on those agents needing it most.

Other Opportunities Increase Employee Satisfaction and Resource Utilization

Continental's reservation centers handle seasonal calls for a limited number of third parties. Call volumes for the airline generally run higher during the summer months and are the lowest at the end of the fourth quarter. This meshes very nicely with a retail situation where call volumes peak during the fourth quarter. By accepting and processing third-party calls, Continental not only utilizes its facilities and earns additional revenue, but also provides variety for the agents—a very popular feature!

"With the exception of technical service, there are not many other types of calls that a trained reservation agent cannot handle," Goodwin states. Over the past few years, Continental agents have handled calls for Houston-based Ashford.com, which sells luxury goods including diamonds and expensive jewelry over the Internet, and Eurail, the European railroad and shipping joint venture, among others. The variety this provides has proven very popular with agents, as has the additional training opportunities for agents to further develop skills and advance in the organization.

STATE FARM INSURANCE

State Farm Insurance, the US' largest property and casualty insurer, as well as its largest insurer of automobiles and homes, enforces approximately 73 million policies through a network of more than 16,000 local agents, 900 claims service centers, and four response service centers. State Farm also claims leadership in terms of customer satisfaction, as measured by the American Customer Satisfaction Index and J.D. Power and Associates' 2001 National Auto Insurance Study(SM). Service, the driving factor for State Farm, is evidenced by the 80,000 employees' firm commitment to delivering "Good Neighbor" service—"being there when and where we're needed to do what we've promised."

State Farm's good neighbor promise was a key driver behind the establishment of its after-hours response centers. For 80 years, State Farm relied exclusively on its network of local insurance agents and claim professionals to service customers. Beginning in 1997, State Farm introduced after-hours service through four response centers located in Bloomington Illinois, Woodbury Minnesota, Jacksonville Florida, and El Paso Texas.

Adding after-hours service added more convenience for customers in their dealings with State Farm. Customers calling after normal business hours are given the option to leave voice mail messages for agents or to transfer to the response center, making it easier to extend coverage for customers purchasing a new vehicle over the weekend or to report an accident.

The response centers employ 1,000-plus customer service representatives—about 250 at each location. Representatives handle phone calls and respond to e-mail inquiries for all State Farm insurance products: auto, homeowners, life, and health.

State Farm's response centers handled about 6 million inbound calls in 2001. Call volume has increased steadily every year since the centers were opened. Center representatives handle several transactions for customers.

Determining Customer Value

The response centers are still relatively new, so management continues to actively determine the services that customers expect and value. State Farm gauges customer satisfaction through numerous telephone and mail surveys, and additionally, management places a strong emphasis on call quality at the centers. To encourage buy-in to the quality process, State Farm involves representatives in the development and implementation of the quality assurance process. State Farm discovered that employee involvement inspires higher productivity and reduces turnover. Management also conducts focus groups with representatives to better understand the front-line perspective and to gather customer feedback. As a result, representatives can focus on the behaviors most valued by State Farm customers. This ensures responsive service and increased customer satisfaction.

People Processes

State Farm's success in customer satisfaction directly correlates to the response centers' front-line representatives. Each call center location is responsible for hiring and training its own representatives. Finding the right people to staff the response centers requires careful screening to identify those with strong oral communication and PC skills.

While State Farm uses traditional channels to attract new employees, word-of-mouth has been the most successful. State Farm feels that its best employees are those who have referred by existing employees. Representatives working in the response center can leverage their experience and knowledge into other positions within the company. Many employees come to the call center and learn the insurance industry and the basics of customer relations with the hopes of spending their entire careers at State Farm.

Once hired, new employees spend several weeks in a combination of classroom training and job shadowing. The formal classroom training teaches the basic technical skills and soft skills, while blending in company culture. Job shadowing helps new employees gain hands-on experience under the one-on-one guidance of an experienced representative. This eases the new employee onto the floor, builds confidence, while reinforcing techniques and knowledge gained during formal training.

Once on the floor, a representative is assigned to a team that typically ranges from 12 to 15 representatives, with one team leader. The centers are staffed with a mixture of full- and part-time employees. Because the centers handle off-hour calls, peaks occur after individual State Farm local offices close, before they open in the morning, and on Saturdays. Management relies on part-time labor to cover these peaks.

Call Quality

Monitoring call quality is the charter for State Farm's Quality Assurance group. In general, employees are monitored several times per month, with team leaders very involved and active in the monitoring process. In addition to call quality, management also measures schedule adherence and availability.

State Farm's ongoing employee training program sharpens skills while emphasizing its focus on excellent customer service. In addition, State Farm believes that employee development is a critical determinant of employee satisfaction. Trainers work with team leaders and the QA group to identify individual training needs. Training builds proficiency in customer service and the insurance industry. Employees can take advantage of the opportunity to move into technical support, quality monitoring, and supervision. The established career path and employee development focus have proven instrumental in improving employee satisfaction and virtually eliminating turnover.

Technology Supports Excellent Service

Technology's efficient deployment supports and enhances service in the response centers. Automated quality monitoring technology enables the QA group to effectively monitor employees more often than was previously possible. While there is nothing novel about the technology used, State Farm tries to make sure the technology truly supports customers' needs.

In a relatively short time, State Farm has enhanced its highly successful 80 year-old network of local agents through the addition of after-hours support for customers and agents. Customer research, employee participation, appropriate use of technology, and balanced performance measurement resulted in the successful deployment of the centers.

Treating customers with courtesy and letting them know that they are important is foremost. While State Farm watches its average handle time and so forth, it concentrates more on treating each and every customer very kindly and meeting their particular needs.

Customer satisfaction survey results confirm State Farm's success in providing customer service that is "in tune" with customer expectations. State Farm's after-hours response centers now guarantee that good neighbor service is available 24 hours a day, every day of the year.

DELL COMPUTER

The world's number one computer systems company, Dell consistently tops customer satisfaction charts. Dell is one of the world's largest computer system companies and on the Fortune 500 list. Technology Business Research ranks Dell as tops in overall customer satisfaction in all categories over a five-year period. Dell's greatest strengths are cited in the on-time delivery, price performance, and cost of ownership. Dell has topped the American Customer Satisfaction Index for several years as well.

With 100 percent of sales originating through customer contact with the Company, Dell Computer's success directly correlates to its ability to provide not only the right product at the optimum price, but also to guarantee that its customers receive the technological support needed to ensure a successful computer experience. With over 1 million incoming calls per month in the Americas and several million calls per month worldwide, Dell's technology support operations lies at the heart of the customer service operations.

Two separate technical support organizations cater to the different needs of consumers and business customers. With a worldwide coverage, centers are established to serve different cultures, languages, and expectations. Over 5,000 agents serve customers in ten technical support centers worldwide. Four centers are located in the U.S.—Nashville; Twin Falls, Idaho; and two centers in Round Rock, Texas, with others in Mexico, the UK, France, Ireland, and Penang. Dell does outsource some services to partners; for instance, the Consumer division utilizes Convergys for peaks in volume and specialty items. Dell's Round Rock center's 1,400 agents focus on business accounts, supporting enterprise, business platforms, servers, and workstations.

Dell technical support agents assist customers with installation of hardware and OS, implementations, and systems support. Support for third-party applications rests with the vendors, not Dell itself. With order status information available on the web, customers also have access to a great deal of information before calling into the center.

Dell's IVR technology enables customers to get into the correct queue, whether the concern is technical support, order status, or problem part dispatch. Although Dell used to play commercials for customers while waiting in a queue, customer response was negative, with an increase in abandons. Dell's goal for abandoned calls is five-to-six percent. ASA time is in the one-to-two minute time frame, with some queues recording times of less than 60 seconds.

Dell maintains a sophisticated database of customer information designed to enhance support services when needed. CSR's have all individual customer details on hand.

Automate Everything

Dell's philosophy—very technology-oriented and "automate everything"—is designed to save time for both the customer and the company. "Quality" is the gold standard for the consumer experience. Dell's main philosophy centers on the customer experience and quality. Metrics follow. While most companies measure the number of calls handled and how fast agents answer the calls, Dell no longer does.

Dell's web-based interface enables technicians to view their metrics continuously, so that they are aware of their performance at any given time. In addition, Dell sends a web-based survey to customers for each call handled by technical support. The results from the 20 percent responding to the survey provide additional feedback to the reps.

Dell's mission is customer-driven, placing great emphasis on consumer-feedback, not the management team pushing directives. Dell's quality philosophy aims for accurate diagnosis of customer support issues. "First time resolution" is Dell's key metric, recognizing that this may translate to longer call handle times. Reps are told 'Do not hang up that phone until the problem is solved. That is how you will be rated'. Technical support reps' incentive programs revolve around this first-call resolution. In addition, the company examines service time, average call times, and wrap times—not individually, but organizationally—as a basis for coaching. Dell also manages wrap time very closely to make sure it is minimal, and hold time (how long the customer is on hold) from a customer experience point of view—customers don't want to be put on hold.

Chat Technology Enhances Service

Recently, Dell incorporated chat technology center-wide to both reduce hold time and to improve communications. Level 1 (L1), entry-level techs can participate in internal chat with experts (Level 2's) while talking with a customer on the phone. Level 2 (L2) techs can assist up to five L1 techs simultaneously. The results have been exceptional, helping with resolution, bringing down our service time, and making everyone work smarter. Previously, if a tech encountered a problem that he/she could not resolve, the customer would be placed on hold while the tech entered an internal queue to wait for an experienced (L2) rep to help resolve the issue. Chat enables a L2 to multitask in assisting four or five reps simultaneously, eliminating most of the hold time. It's also transparent to the customer—techs can leverage this communications channel to provide better information to the customer and at the same time appear to be very well informed.

First call resolution goal is 85 percent, with an 83 percent performance typically. Dell always ups its goals as soon as they get near them.

At the conclusion of a call, the customer is immediately queried as to whether or not the problem was resolved and if he/she was happy with the call. If there is a problem with technical resolution, the customer immediately goes to a high-level rep; if there is a problem with the process, the customer immediately transfers to a manager. The goal is to catch and solve the problem before the customer hangs up—don't let the customer leave unhappy.

An additional metric that Dell tracks is the occupancy rate, with a goal of 80 percent. With an average of 125 calls per week per tech, Dell watches the metrics to ensure that staffing is sufficient to serve the customers' needs. Other factors examined are the repeat-dispatch rate, service time, and attendance.

Web-based Customer Support

Dell inaugurated an online resource center, the Dell Solution Center, where customers can access Dell's knowledge base to resolve problems or access support.dell.com. The Dell Solution Center also allows customers to view their product manuals or participate in online training in using their computer. The Resolution Assistant provides live chat sessions with support techs, and with the customer's permission, the tech can view information regarding the hardware and software installed on his/her PC. When Resolution Assistant was introduced in 1998, 28 percent of Dell's WebPC desktop owners utilized Resolution Assistant. Seventy percent of this group subsequently used the Resolution Assistant the next time they needed support. Successful first experiences on the web site leads to return visits.

Customer Feedback Drives Changes

When Dell began conducting surveys, obtaining real-time customer feedback, it found that the IVR menus were too cumbersome. The Company responded immediately, reducing menuing from three minutes to one. A second major complaint was the length of hold time, and Dell alleviated that by initiating training that enables the tech to keep the customer engaged while getting technical expertise behind the scenes (chat technology).

Over the past few years, Dell's high-level techs have been calling customers the following day to ask if the problem was resolved and if the customer was satisfied. If the customer expresses dissatisfaction, the tech's goal is to solve the problem immediately. This has been a big boost to satisfaction. Customers don't expect a high-level tech to call them the next day.

Employee Recruitment and Training

Dell seeks candidates with technical troubleshooting experience—not necessarily college degrees, but tech school grads, "someone who really understands the hands-on versus theory." In addition, the recruiting staff looks for certifications, hardware experience, and customer service skills. An applicant will interview with three levels of management at Dell. The recruiting staff and a manager handle the "people skills" interview and an L2 conducts the technical interview. Dell uses a temp-to-perm strategy for new L1 recruits. They go through training, and after the first month take tests. They are converted to permanent after three months. L2's are recruited both internally and externally, going after specific talent—not a temp strategy.

Dell's specialized tech training staff consists of 40-to-50 people who conduct sessions for new hires, new products, ongoing training, and computer-based training. Through self-evaluation, the training team constantly changes its presentation, moving to more hands-on and self-paced training activities. By enabling trainees to move at a self-paced rate the first week, everyone comes into the classroom training on the same level. The trainees take a knowledge test as he/she comes into the training session to assess needs.

Hands on training is considered very important. Techs must know what the systems look like and how to install things, just as the customer experiences. Dell also brings in suppliers and vendors to show techs how certain applications work so that they can better understand. Vendors also set up rooms with their technologies to demonstrate how broken cards would function and how they would look, for instance. Techs can also explore and talk with vendors.

Respect, Advancement, Pay Lead to Satisfaction

Dell enjoys a low turnover rate, thanks in part to its philosophy of reinforcing positive metrics with individual awards, team rewards, and cash-back bonuses. The best tech performance each quarter earns cash rewards and an annual drawing amongst all quarterly winners awards $15,000 to the winning tech.

Technical reps enjoy a career path similar to management and engineering. From L1 to L2, techs can also progress to the gold support level for clients and to enterprise gold—there are six levels equivalent to Dell's senior management. Dell also supports continuing education for techs through industry standard certification. Teams comprised of 20 techs per manager work in the same area, with one or two L2's assigned per team. L2's work closely with the L1's, making the next-day callbacks, and coaching the L1's as necessary. L2 bonus incentives are tied to the team's success. Dells teams are specialized by product line or customer. For instance, Dell's tech teams function as help desks for many major corporate clients. Assigned a special incoming number, those calling the special number are unaware that the tech assisting them is a Dell employee, not a corporate tech employee.

An open-door policy and an open environment add to employee satisfaction. Each tech has his or her own walled office—this is quite different from most tech support call centers. Additionally, since techs talk to Dell's most important customers each day, they are rewarded with respect and an excellent pay scale.

Occasionally Dell will train a few techs for highly skilled processes in order to better serve a particular client. When the company doesn't want to train the entire floor on a special process, it solves the issue with specialization.

Customer Satisfaction: the Ultimate Metric

Besides the end-of-phone-call survey and daily transactional surveys, Dell tracks escalations—letters to executives or Michael Dell himself—and handles each instance. By measuring customer satisfaction through metrics in place, Dell has the means to solve problems quickly and efficiently. The Dell approach is: if you measure it, put metrics in place, give incentives to make aggressive goals, you get there. Dell's culture revolves around customer satisfaction. Initiatives developed by Michael Dell are published on internal reader boards, bulletin boards—"everywhere, everything is tied to the customer experience. Financial incentives depend upon meeting customer satisfaction metrics.

FEDEX

"Our goal is to deliver the perfect customer experience every time – one that's seamless, convenient and efficient." Widely acknowledged as a great place to work, a technology leader, and one of the world's most admired and respected companies, FedEx Corporation ("FedEx") is the leading provider of transportation, e-commerce, and supply chain management services. Topping the American Customer Satisfaction Index in its category for many years, FedEx consistently ranks as one of Fortune's "Most Admired Companies" and "Best Companies to Work For." The FedEx web site, acclaimed for its speed, ease of use, and customer-focused features, gained Web Marketing Associations' "Best Transportation Web Site", as well as NetMarketing's "Top Business-to-Business Web Site." The first company to win the Malcolm Baldrige National Quality Award in the service category, FedEx Express also was the first global express transportation company to receive simultaneous system-wide ISO certification.

A world leader in technology, FedEx sets the industry standard for efficiency and customer service with its commitment to customer-driven technology. As stated by CEO, president and founder Fred Smith in 1979, "The information about a package is as important as the delivery of the package itself." This philosophy remains the Company's driving force. The Company's development of door-to-door package tracking capability simultaneously raised the satisfaction standard and changed the transportation industry.

From the beginning, FedEx has led the industry in new service introductions. The first company dedicated to overnight package delivery, FedEx also led the way in introducing the air express industry to the concept of the Overnight Letter, the 10:30am Next Day air, Saturday delivery, and on and on. The company puts its customers first; forging strong relationships and developing innovative solutions based on customer need. FedEx Express opened its first customer service center in 1978 and now manages 17 centers in the U.S. and 43 internationally. All domestic centers are linked to create one virtual center. FedEx operates a 24/7 call center operation with approximately 4,300 agents handling more than 640,000 calls each day.

FedEx customer service representatives arrange for package pick-up, track the delivery status of packages, offer rate information, and suggest packaging. FedEx also offers self-service options through its Integrated Voice Response technology and web site. FedEx's IVR handles more than 20 percent of total call volume. Web site transactions account for almost two-thirds of the five million shipments FedEx delivers daily.

Enhancing the Customer Experience

FedEx customer service operations have benefited significantly from advanced telephone and computer automation. Several years ago, FedEx implemented CTI (Computer-Telephony Integration) screen-pops to match incoming callers with their account information. This alone produced a significant savings in call handle time as well as improved customer satisfaction. In fact, CTI provided the foundation for FedEx's Customer Relationship Management system (CRM). FedEx began building customer profiles following its CTI implementation—tracking the reasons for each call by individual caller. This enabled management to analyze call patterns, isolate root causes, and ultimately improve customer service levels.

The company continues to pursue new technologies designed to enhance the customer experience. Mike Feehan, managing director of Customer Service Strategic Planning and Engineering is responsible for the group that identifies improvement opportunities within the customer service centers. "It's all about creating a better customer experience," states Feehan.

"We're trying to better understand what our customers want and provide our agents with the right tools so they can deliver up to these expectations."

FedEx recently replaced all its workstations, servers, LANs, and ACDs (Automatic Call Distributors) in its customer service centers to improve throughput. The IVR was recently upgraded to support natural language speech recognition. In addition, the IVR can support the completion of multiple transactions on the same call. For instance, a caller can schedule a pickup, determine the shipment cost, and even order more supplies—all without the assistance of a representative and with only one phone call.

FedEx just completed the global implementation of a new Clarify CRM system. So far, management agent feedback is positive regarding the new CRM desktop—it's much easier to use, more intuitive, provides access to more information, and requires less training. "CRM makes it easier to have intelligent discussions with customers," adds Feehan. "We'll be able to provide better personalized service and better tailor interactions for customers." The new system gives representatives a comprehensive view of every customer, detailing each customer's needs, and suggesting services that might meet those needs.

A newly deployed knowledge management (KM) application provides easy access to all of FedEx's internal documentation—rate charts, job aids, daily communications, process descriptions, technical references. Customer service representatives access the KM system through the customer service representative home page. "We're just beginning to appreciate what KM can do and learning how to take advantage of it," adds Feehan. "We continually enhance the desktop to make interaction better between the agent and the customer," adds Feehan. CTI, CRM, and KM are all ways of doing just that. FedEx plans to add skill-based routing in the near future. It is also exploring adaptive selling tools for representatives.

FedEx utilizes web-based training for new hires as well as for recurrent training needs. For instance, FedEx developed a web-based training module to demonstrate how to handle a "pick-up" call. The training module walks the agent through a typical pick-up call, demonstrating all the screens and commands necessary to schedule a pickup. The module concludes with a post-training test. This makes it much easier to keep representatives current on new processes and products.

Technologies supporting FedEx Express and FedEx Ground are linked to provide seamless and consistent service, regardless of the chosen package delivery speed. The continued emphasis on the pursuit of new and enhanced technologies continues to pay off for FedEx. Customers are happy, employees are happy, and so are stockholders. "The new technology is a winner. It's making the agent's job much easier and as a result, employee satisfaction is higher," comments Feehan.

Perfecting the People Processes

All FedEx customer service representatives are hired into the company as full-time employees. "We hire permanent employees, not temps. We have a big investment in employees once they come on board," states Feehan. After four weeks of training, newly hired employees graduate into a "nest." For several weeks after the training program, new employees sit with experienced representatives as they handle their first calls. This creates a support system for the new employee—the "buddy rep" is right there if a question arises.

"We're constantly looking at turnover, how we hire, who we hire, and whether or not expectations match," states Feehan. As a result, management is continually identifying ways to improve processes and systems to reduce turnover. In addition, all employees are interviewed when they leave the company.

In terms of performance, FedEx uses a "Best Practice Scorecard" to review performance and, ultimately, determine compensation. The scorecard features a series of metrics that represent a balanced collection of efficiency, effectiveness, and quality measures designed to emphasize the correct actions. Representatives are scored on each measure and an overall score is calculated based on weightings. A similar scorecard is calculated for each team and each customer service center. Scorecards are posted weekly for all employees to view. The scorecards determine merit increases as well as a "pay for performance" lump sump payout. As part of the quality component of the scorecard, managers and FedEx's Call Quality Group monitor representatives regularly. Feedback provided to representatives comes in a positive manner, geared towards personal development and improvement.

Measured daily, customer satisfaction is charted through daily calls to randomly selected customers, with a certain percentage of all customers contacted to determine the customer experience. The survey questions each caller about his overall experience as well as specific attributes of the FedEx representative's performance. In total, each call measures 15 different attributes. The accumulated results are reported back to each team manager as feedback. While the sample isn't large enough to support valid results for every representative, the feedback is quite useful on the team level. Satisfaction results can identify problems for follow-up observation. It also provides a sense of the customer's experience in a way in which representatives can relate.

FedEx's employee-centric culture is an integral part of its customer satisfaction strategy. If a company takes care of the needs of its people, its people will take care of its customers.

The Value Proposition

According to Feehan, FedEx's overall customer service strategy provides a focus on the issues that the customer might experience and seeks solutions from a product standpoint, as well as issues and problems. In addition, FedEx offers customers multi-channel service options, leaving the choice to the customer. "We emphasize self-help as much as possible where it makes sense," states Feehan. "We work to provide customers with different channels to complete their transaction, whether it's the IVR, fedex.com, or the call center."

Recognized as a technology pioneer from the beginning, FedEx's Executive Vice President and CIO Rob Carter comments, "Having a technology advantage is very important to FedEx, so we're always looking for innovative new technology that can speed up a process, eliminate rework, and improve quality." FedEx is clearly moving beyond customer focus, because merely knowing your customers no longer suffices in today's competitive market. FedEx strives to create maximum value for customers by listening to them, anticipating their needs and preferences, and continually developing and delivering new and valuable products and services.

LOUISVILLE GAS AND ELECTRIC

LG&E Energy includes 3 utilities: Louisville Gas and Electric (LG&E) , Kentucky Utilities (KU), and Old Dominion Power (ODP). LG&E and KU between them serve most of the State of Kentucky. ODP is a smaller utility in Virginia. Louisville Gas and Electric has topped its category in the JD Power customer satisfaction awards in 4 out of the past 5 years.

Call Centers

LG&E's call centers handle 2.1 million calls per year and the IVR system 0.5 million. All calls are recorded and most also have the screen captures recorded. Calls are archived for 3 years. Ten calls per CSR per day are monitored, and coaching provided. CSRs have individual feedback and are given a lot of reward and recognition for good performance. They are also paid extra based on skills and receive bonuses for being called-in to cover for absences.

For CSR hiring, LG&E uses an agency for pre-screening. This results in much better candidates than before, and the agency guarantees to take back CSRs that don't work out. Some potential hires start as contractors and work alongside employees. Contractors are eligible to become employees when vacancies come up. They are on a 6-month trial period, and then as with all new employees are on trial for the first 6 months of employment. This hiring/trial process is considered by LG&E management to be one of its best practices.

The call center turnover is currently 29%. Half of this is considered desirable as it represents people who haven't worked out. For people who need improvement, personal improvement plans (PIPs) are developed, but if this does not succeed the CSR has to eventually leave. The PIPs included various measures including what other parts of the company think of that person's performance. CSRs obtain regular feedback in this way.

Team meetings enhance morale building. LG&E has 3 call centers, so they pick a slow time and route the calls to 2 of them while the other call center's staff is at the meeting. CSRs also spend time in other parts of the company (the "Shadow Program") to learn about those areas. LG&E does a lot of smaller things like providing lunch for CSRs on Mondays since they are busy, having fun days such as "mint julep days", providing spot rewards such as sports tickets and massages.

New CSRs are coached using the buddy system. Also there is a CSR intranet: "RoboInfo", which links to internal specialists on various topics of interest to CSRs. This year (2004), LG&E is hiring a training expert from Cisco to further improve CSR training.

Business Call Center

LG&E has had a separate call center for business customers for the past 4 years. The company finds that this serves the needs of business customers much better than a combined call center. The company links 2 sites as a virtual business call center. About 10% of total calls are from businesses, and about 10% of CSRs are in this dedicated virtual call center.

Management focuses on 4 areas: customers, employees, cost and process performance, similar to the residential customer call centers. Business customers have a single point of contact and obtain specialized services, especially tailored to builders/developers and property managers/landlords.

The CSRs also handle email, faxes, mail and in-person customer contacts. Faxing is a very popular communication medium for contractors.

The call centers' performance focus is on courtesy, call handling quality and ASA (the last aspect is very important to business customers, and is currently 11 seconds).

Specialists are available to assist the CSRs. These specialists spend 50% of their time on site visits to customers, and have specialized training on tariffs and energy supply contracts. They conduct 5000 business customer visits per year.

Meter reading

LG&E reads some meters for others (Columbus Gas). They transfer a file from the Itron hand helds to Columbus Gas who produce their own bills. This adds about 100,000 meter reads a month to LG&E's 1.2 million monthly reads for its own customers. In addition, there are about 60,000 reads done by various AMR systems. Generally they use AMR for hard to access or dangerous areas.

LG&E obtains a lot of productivity measures from the hand helds, e.g. number of re-reads. Forty-five of its meter readers are employees, while 120 are contractors. The productivity is the same, but employment costs are higher for employees. There is a scorecard every month that is reviewed by senior management, and the performance targets are deliberately set high.

Sharing meter reading technology and systems between LG&E and KU has saved a lot of money: water utilities should consider this re "regionalization".

LG&E uses the ARCs system (Automated Routing Control System) from DB Microware to regularly re-optimize its meter reading.

LG&E has held customer focus groups on metering to discuss ideas, with good results. Ideas have included reading meters quarterly but billing monthly, and customers have said that up to 5% estimated reads is not a problem.

Remittance processing

Recently there is a stronger focus in the eyes of customers on payment alternatives. LG&E has also contracted out the pay by phone options (credit and debit cards) to a 3rd party, Bill Matrix, and there is a convenience fee for most of these options (except electronic checks). LG&E's customers do not balk at paying the convenience fees. Customer satisfaction has increased with multiple payment options, and there has been no change in the bad debt ratio.

Credit and collections

LG&E's bad debt ratio is 0.46% (KU is lower at 0.2%). The company has an aggressive disconnection policy (the very first day allowed by policy), and has a team of 6 people who exhaust in-house collection efforts before turning hard cases over to a collection agency. By this stage, the collection agency only recovers about 17% of debts it chases. The company rarely uses property liens.

Customers who do not give a SS# usually put up a 2-month deposit (cash for residential, LOC or bond for I/C). LG&E also uses a 3rd party credit rating agency, Accurint, which is web-based and will provide a credit and background check for $0.25, with no upfront fee. The PSC governs deposit and disconnect policies (as well as tariffs), which helps with enforcement.

Theft of service is a problem, and the company has 2 people who work at this constantly, e.g. checking on disconnected services that do not reconnect within a short time.

Budget billing did not reduce bad debt until the company made customers "true up" their accounts periodically, then it did.

LG&E runs a scheme whereby it matches $ for $ donations from customers to help poor customers pay their bills.

Meter services

LG&E has had significant savings from mobile dispatch, reducing meter service staff from 57 to 44 FTEs. These employees use mobile data terminals that they take into their trucks in the morning with tasks loaded in, and they upload task status to the network nightly. LG&E does not use GPS as they have not found it necessary (but they acknowledge that other utilities may have uses for GPS such as locating stolen vehicles).

The use of contractors has increased greatly, and is now about 50:50 in field services. Managers feel that such a mix maximizes productivity and quality.

Innovative or Winning Practices

Various functions are between 20-80% outsourced. This practice keeps pressure on both employees and contractors re quality and productivity.

Senior managers receive regular, very detailed performance figures on all aspects of the business.

PUBLIC SERVICE COMPANY OF NEW MEXICO

Public Service Company of New Mexico (PNM) serves 440,000 customers with natural gas (400,000 homes and 40,000 businesses). The PNM gas system is 12,378 miles long. It includes 1,478 miles of pipeline and 10,900 miles of distribution and service pipe.

Up to 2000, PNM was a functional organization operating in a "silo" fashion. In addition, following the takeover of another gas utility, they were suffering with two incompatible legacy CIS systems. Wanting to improve performance, PNM's initial approach was to document and implement approximately 150 standard procedures in 60 days. This didn't work out well, and PNM struggled for 12-18 months before changing the approach to customer-focused culture change.

This culture change from "wires and pipes" to customer relationships was not easy. PNM mentioned "battles" along the way. Many of the senior managers had to be replaced to make the changes happen. This included the VP of customer service and all his direct reports. PNM says their culture was a little stodgy with a "we have always done it this way" approach. Not moving out the people who were not going to change would have been a big mistake: in the past such people had dragged the change effort down. Overall, the most important element for change was support at all levels in the management hierarchy.

Call Centers

CSRs were until a few years ago made to feel incompetent, easily replaced, judged on efficiency only (# calls handled, wrap up time, talk time), and were only spoken to regarding problems. Now, PNM listens to its CSRs and has changed the way it measures their performance to focus on quality customer service and first call resolution, not minimizing call time.

The increase in first call resolution rate has reduced call volume and both customers and CSRs are happier. Reducing CSR frustration has decreased call wrap up time from 126 seconds in 2001 to 44 seconds in 2004, because CSRs are not looking to put off taking the next call. Also as a consequence, average speed of answer has reduced over the same period from 574 to 35 seconds (recently it has spiked to 54 seconds due to a sharp increase in gas prices).

Another contributing factor to better CSR performance is a new bonus scheme (up to 7.5% of salary) based work group performance (customer satisfaction scores, nationwide benchmarking comparisons and group call quality monitoring scores), and corporate performance. Yearly merit raises are on an individual basis (availability, individual call quality monitoring scores, attendance and overall customer satisfaction, plus 2-4 measures at each CSR's discretion (e.g. doing voluntary work, training, making presentations.) Note: CSRs do not belong to a union.

Some time ago, to further improve service levels, 15 new CSR positions were approved, however due to improved performance as above these have not been needed.

Selection of CSRs is seen as an important contributor to better performance. The first interview is by phone. By checking the applicant's phone manner in this way, a lot of unsuitable applicants are weeded out. The next interview is in person with a 3-member interview team. Questions include "what types of volunteer service do you do?" (Indicates desire to help others), "tell us about a time when you bent or broke a rule to help a colleague or customer" (indicates ability to think outside the box to help a customer). Since good potential CSRs are in high demand, once PNM decides to make an offer of employment, it does so quickly, usually making contingent offers within 5 days and sometimes on the spot. The contingency relates to background checks. Vacancies in the call center used to take 54 days to fill, now 15 days.

A new performance coaching program involves monitoring calls, which the CSRs and marketing staff review. The criteria used for rating calls is:- Greeting: 5%- Problem solving: 20%- Transaction execution: 35%- Professional communication: 20%- Account documentation: 15%- Closing: 5%

Front Office

Many customers pay in person so PNM maintains walk in payment offices. Walk-in/call center staff members are managed as a pool, to provide the flexibility to cope with demands for service and to provide cover for absences.

Bonus payments are based on a composite of call center and walk in customer satisfaction, ensuring that staff members are focused on both. Even senior level CSRs take part in job rotation.

Innovative Practices

PNM has reduced end-to-end process time, through process mapping by cross-functional teams led by customer service people in all cases. These teams examine the many variations in the way the same tasks are carried out and this results in a lot of learning. PNM defines key processes to include account management (includes service initiation, account setup, billing and handling billing exceptions), metering (includes processing metering data), credit and collections, handling customer-initiated contacts, handling company-initiated contacts, and carrying out service orders. Process owners resolve anomalies and set standards.

Focus from CEO down is on customer relationships; no longer is focus on "wires and pipes". As an example, the Customer Service Group used to be treated as a cost to be reduced. Now it is seen as leading customer relationships, and has received a lot of funding in recent years for improvements, including call center facility and technology upgrades.

Customer communication and media releases get passed by the customer service group first, so that the CSRs know in advance what their customers are seeing.

Customer service is involved in rate filings: "how will this rate design be received by customers?" Feedback from customer service has resulted in changes to rate filings, e.g. a recent large rate increase was delayed until proper customer communications had been made. Stakeholders are also involved in rate structure design.

PNM recently ran a "Beyond Customer Care" program where selected CSRs spent 17 days with other departments that they interact with, learning about what these other departments do. Communication between departments makes a lot more sense now.

Additional revenue sources are being actively pursued, including charging fees for doing research for attorneys, and becoming a 3rd party verifying agent for phone, cable and newspaper companies. In addition, an upcoming call center technology upgrade will allow services such as wake up calls (especially important re power outages), automated group calling for others (e.g. non-profits) and working with the Department of Homeland Security re mass notifications.

PNM considered CRM technology but decided it was not needed since they have IVR, web and good call center technology.

PPL ELECTRIC UTILTIES

PPL Electric Utilities provides electricity delivery to 130,000 accounts in eastern and central Pennsylvania and markets wholesale energy in the US and Canada. In July 2002, J.D. Power and Associates named PPL Electric Utilities first among electric companies in the East, according to its 2002 Electric Utility Residential Customer Satisfaction Study. PPL Electric Utilities (PPL EU) garnered the top score among 15 eastern companies in all five measures of satisfaction: image, power quality and reliability, price and value, billing and payment, and customer service. Recognition for superb customer service is nothing unusual for this utility, which is a subsidiary of Allentown-based PPL Corporation. This is the fifth J.D. Power and Associates award and for the second year in a row, PPL placed first in the American Customer Satisfaction Index in its category.

As part of PPL's commitment to Electric Choice in Pennsylvania, PPL plans to offer customers new and different options for electric service. Automated meters will enable PPL to test concepts and make future alternative pricing plans possible. PPL Electric Utilities also is using the two-way communications capability of its automated meters to help restore electric service more efficiently after major storms. By sending an electronic signal to meters, the company is able to determine whether power is flowing through them.

Call Centers

Dedication to customer service drives the company's commitment to quality in its contact center that fields over 2 million calls each year. With approximately 1.3 million customers in 29 eastern and central Pennsylvania counties, PPL Electric Utilities' agents handle 1.75 million of these calls, while approximately 250,000 calls are served within the Interactive Voice Response (IVR) system.

State-of-the-art technology, well trained and customer oriented employees, and innovative practices are the tools employed to consistently deliver award-winning customer service.

Contact Center Technologies Provide Convenience to Customers

One virtual network connects three call centers located in Scranton, Allentown, and Hazleton. The three centers share one common Automated Call Director (ACD) and one IVR. Responsible for customer care, billing, collection, new/upgraded service, outage reporting and dispatch, the call centers employ approximately 200 agents who not only handle calls, but also perform back office work. Part-time agents help provide coverage during lunches and breaks.

Through the IVR system, customers check account balances, report outages, check outage status, and connect to corporate employees through a switchboard application. In February 2002, PPL EU implemented a new IVR and ACD platform. The new system includes speech recognition, text to speech, computer telephony integration (CTI) and easily integrates with the skills-based routing application. The IVR directs calls to agents in four distinct service groups' outage reporting or electrical emergencies, connects/disconnects or new service, billing investigations, and collection. Business customers utilize a dedicated toll-free number to dial in, bypassing the IVR and connecting directly to business specialists.

Screening Process Identifies Best Candidates

"Our people are exceptional," states David Ling, Director of Customer Operations. "Our people have a real penchant for satisfying customers. That is our secret weapon - our people." Continues Ling, "We do a lot of screening and testing during the interview process. We test for attitude and behavior using a standardized test, and we pre-screen for call center and customer service experience." During the interview, candidates answer a standard script of behavioral questions to determine how the candidate would handle certain situations. "We even do role playing", comments Ling. Role playing is conducted during the interview to develop a feel for the candidate's ability to answer customer concerns in real time.

PPL Electric Utilities primarily uses instructor-led training for their agents. The training material is available on-line for reference by all agents and supervisors at any time. "We tap into the skill and job knowledge of our top performing agents to facilitate the training. They assist our instructors during classes and work with the curriculum developer on the training materials. This way, we are assured that our training material flows well with how the job is done. It also gives our agents the opportunity to provide input into the training program", states Ling.

New agents are trained on specific call types. After receiving the training, they begin taking calls in the classroom with the training assistant as their coach. When they have mastered the call type, training will begin for the next module or new skill area. New employees are first trained to handle collection-related calls and connect/disconnect calls. It takes approximately 4 months to complete the training program; however new agents take their first call after 56 hours of training.

The company promotes from within as much as possible, which improves employee retention. The Candidate Assessment Program guides potential supervisory candidates through the process. A screening test qualifies agents as potential supervisory candidates, making them eligible for supervisory training. When a supervisor position opens up, a qualified pool of candidates stand ready to step in.

The contact center monitors calls, conducts random weekly customer satisfaction surveys with customers that have called that week, and subscribes to external studies. In addition, the company conducts its own broad-based customer satisfaction research quarterly.

Management meets with agents regularly, discussing corporate goals, customer satisfaction results, and individual performance, including call quality results.

Prior to the implementation of the new eQuality monitoring system, it monitored approximately one call per employee per quarter. The new technology enables the recording of one or two calls per agent per day, including screen navigation as well as voice. The quality monitoring group assists supervisors in administering the program. In addition, supervisors have the ability to plug in to assist agents with calls or to coach an employee with a particular skill. Supervisors can plug into agent calls from their desks.

Each day in the contact center begins with a "stand up" meeting with all supervisors. Why stand up? "Everybody stands," Ling says, "and it keeps the meeting from lasting more than 15 minutes. We go over the daily plan, what work needs to be done, which group is doing what work, available staffing levels for the day and expected call volume." If information needs to be shared with agents, a "Please Share" segment is added.

Management is constantly on the lookout for ways recognize superior performance within the call center. "We find a lot of reasons to celebrate", states Ling. "Pizza days, buffet lunches, and bagels and fruit mornings are always popular."

Dedicated Group Caters to the Business Customer

In 1998, the company created the Industrial and Commercial Services (ICS) group, providing customer care services for mid-to-large business customers. ICS handles approximately 18,000 calls per year, with employees based in the Hazelton and Allentown offices. ICS employees are specialists that function like phone based key account representatives to serve nearly 25,000 accounts.

ICS strives to answer 100 percent of its calls with no wait time. There is no IVR for business customers, there is a limited queue, and there is generally very little wait. Business customers may opt to leave a voice mail message if an ICS agent is not immediately available. The company's research indicates that business customers are more attuned to voice mail than waiting in a queue. Not surprisingly, this same research noted that business customers prefer ICS specialists to answer their calls directly.

ICS also conducts customer satisfaction transactional surveys, which provides customer satisfaction results for each specialist. The scores resulting from these surveys, in part, determine the ICS representatives' performance and pay.

Meter Reading

PPL has implemented an AMR system covering virtually every customer. It is one of the first large-scale implementations of Power Line Communication (PLC) technology. The system sends meter reads and has 2-way communication with meters via power lines.

PPL sees 3 types of benefit: tangible, intangible and future. The business case was justified on tangible benefits only, which were: 1) elimination of manual meter reading, i.e. labor, vehicles, etc.; 2) elimination of estimated reads which reduced the call center workload considerably; 3) ability for high bill complaints to be resolved by the CSR while the customer is on the phone, as the CSR can ping the meter and also examine daily readings. This saves many special meter reads; 4) improved accuracy of large meter reads and elimination of manual checks; 5) outage diagnosis can be carried out without sending a crew, i.e. PPL can establish from the call center whether it is the utility's or the customer's side of the meter that has the problem; 6) the utility can easily confirm whether all customers are back on line after an outage by testing all the meters in the affected area; 7) short term meter maintenance costs are reduced because all meters were recalibrated or replaced during the system implementation.

Intangible benefits include: 1) reduced theft of service, which is the biggest intangible benefit by far. National estimates are that 1-3% of electricity is being stolen, however by definition the amount at any one utility is unknown, so even though it was known that the AMR would reduce this problem, no estimate was used to justify the system; 2) capacity to read meters hourly; 3) meters can flag if they have been tampered with; 4) data mining of customers' energy use patterns; 5) improved metering accuracy; 6) better customer service and higher satisfaction, leading also to more favorable image with the PUC.

Future benefits are expected to include: 1) more optimized preventative maintenance, ability to make more precise decisions re periodic maintenance than just following fixed maintenance schedules; 2) possible support to more tailored tariffs if time of use charging can be increased; 3) more granular understanding of usage could improve wholesale energy purchasing decisions; and 4) more accurate financial reporting, less estimated usage.

The system is working well. PPL feels that the system would be especially applicable to medium-sized munis and co-ops in semi-rural areas. Change management was easier than other projects such as CIS, WMS and Y2K, because the AMR system made CSR's jobs easier, it is a specific application without many of the integration problems of the other systems, and the AMR system was rolled out in phases.

Other Noteworthy Practices

Web self-service joins the IVR as another focus of improvement efforts. Through www.pplweb.com, customers can access information regarding new accounts, view and pay bills, prepare a home energy audit, and obtain information on energy-saving appliances or home modifications.

In the fall of 2001, PPL Electric Utilities introduced Rapid Service Request, a Web-based method for builders and contractors to submit requests for electric service to new homes. It also provides a quick and easy way to check on the status of service requests. Rapid Service Request also includes a convenient means to update common job status information and to provide information about issues or concerns related to the installation of electric service.

Balanced Scorecard Focuses Employees

PPL Electric Utilities strives for balanced performance in four key areas: customer service, employee performance and satisfaction, operational performance, and financial performance. The company employs a wide variety of initiatives in its pursuit of enhancing performance in each of these four areas.

In November 2001, the company introduced an annual employee satisfaction survey. In addition, it continues limited surveying between these formal annual surveys. Electronic messaging is also used to send announcements and to share information with all employees at the same time.

WASHINGTON SUBURBAN SANITARY COMMISSION

Washington Suburban is among the 10 largest water and wastewater utilities in the country, serving 420,000 customer accounts through a system of more than 10,000 miles of water and sewer mains. More than 1,500 WSSC employees provide water and sewer services to 1.6 million residents in a 1,000 square mile area.

Call Center

The WSSC's IVR provides more information and options than many others, such as bill and payment information, extend payment due date, enter a meter reading, directions and office hours, make a change to property ownership.

Combined emergency and non-emergency call centers leads to better communication with customers as well as better utilization of employees.

Intensive training is provided to new customer service representatives. The training consists of two phases each ending with an exam. There is six weeks of classroom study, five weeks of on-line training and five field trips.

CSR's performance is monitored against standards (availability on-line, time spent in work mode, length of talk time), using quality monitoring software "Click to Coach", which can mirror what is said and what screens are being accessed in what order to help determine processing efficiency.

All payment options can be handled by CSRs. These are mail, direct debit, on-line credit cards, check by phone, 3rd party agents e.g. banks, lock box, and walk up/drive through.

Meter Reading

Almost all small meters, both inside and outside settings, can be read by touching the meter reading device to the remote reading device. New meters are all AMR capable (Schlumberger/Neptune Proread meters since 1995.)

The WSSC has a meter reading quality index system that tracks meter reader errors. Meter reader performance measures include: Routes completed points - points are awarded for each route completed or scheduled paid vacation. Error deduction – points are deducted for each error that is received during the month. The number of errors allowed per month is a 0.25% error rate, which equates to less than 15 errors per month. Job deduction – points are deducted for the number of outside no reads. Inside read points – points are added for the number of inside reads obtained. Customer compliment points – points are added for each compliment received. Customer complaint deduction – points are deducted for each customer complaint. Free form comments points – points are awarded for valid free form comments.

Meter Readers are located in two different depots but report to one centralized management structure. Personnel are shifted between depots as needed. All meter readers are trained to read every type of meter (small meters 5/8" to 1" and large meters 1.5" and above located inside and outside in pits or vaults).

The WSSC re-balances meter routes annually using Route Management Software.

Meter Maintenance and Related Field Services

Replacement criteria and assumptions for small meters are re-evaluated every three years, especially with regard to the latest data on age versus consumption approaches for meter replacements.

Large meters are tested more frequently and testing is based on consumption.

There is a current study on the effectiveness of touch meter reading (e.g. over time, some meters are becoming obstructed if they do not need to have the pit opened).

Reports are automatically generated on high and low reads.

A new meter test bench is capable of uploading meter test results to the mainframe.

There are penalties for anyone installing a meter backwards, or disconnecting the wrong meter. Since these disciplinary procedures were introduced, the level of such errors has dropped dramatically.

Field representatives are cross-trained so that they can work in dedicated areas. Field representatives have either cell phones or Nextel phones. Nextel was invaluable on 9/11.

Other Innovative Practices

The customer service and meter service organizations were combined in 2000 with the distribution and collection systems maintenance group. A single Customer Care Team is responsible for all of these functions. That is, all field personnel and customer care personnel report to the same manager.

There is a Customer Care Strategic Planning Group, which focuses on creating alignment with the corporate strategic plan, designs and delivers training courses, and coordinates special projects such as Automatic Meter Reading, Optimum Meter Age Study, and Flexible Worker skill tracking.

There is a bounty program, where any employee that finds lost revenue (e.g. unknown meters) receives a share of that revenue.

Customers are charged for the cost of a requested meter test if the meter performs within tolerance.

Customers receive a $25 credit for missed appointments or other “unpleasant experiences” with the WSSC.

Laptops in the field provide quick access to information without having to request assistance from the Call Center staff. In addition, the field staff uses the laptops to close out work orders, which provides timely information to the Call Center staff. The fact that these personnel are all within the Customer Care Team also has had a marked benefit in co-operation.

There is a water fund to assist people who have difficulty paying their water bills, administered by the Salvation Army. Customers may contribute to this fund, and employees also raise money for it.

Future Plans

Electronic bill presentment and payment will provide customer ability to view their bill and pay it on-line through a service such as Check Free.

Automatic meter reading (AMR). The WSSC has tested 6 vendor systems and has a strategy to start implementation within a few months on the monthly read meters. These meters represent 1% of the meters that account for at least 40% of the annual revenue.

Unitize Measuring Elements (UME), which provides quick / accurate removal and replacement of the large meter measuring elements make it safer for personnel to change out the measuring elements in vaults.

The WSSC intends offering meter testing services to other agencies, to provide water and sewer service applications on-line, and offer "Leakguard" as an option, whereby customers would pay a low monthly fee and the WSSC would be responsible for any repairs on customers' property up to the meter.

Claims tracking software on line with more details on the status of claims, reducing a lot of paperwork

YARRA VALLEY WATER

Yarra Valley Water (YVW) provides water supply and sewerage services to 1.5 million people who live and work in the Yarra River catchment area of Melbourne, Australia. YVW is owned by the State Government of Victoria and operate commercially under a Board of Directors appointed by "shareholders" (in this case, government ministers), and is subject to Corporations Law. YVW is subject to regulation by the Essential Services Commission and makes payments to the Victorian Government equivalent to the income tax and sales tax, which would be payable if the Company were not State owned. YVW's directors adhere to corporate governance practices based on the Australian Stock Exchange Listing Rules, even though it is not a listed public company. The company has adopted the Code of Conduct issued by the Australian Institute of Company Directors.

Call Center

YVW has a single integrated call center handling all call types (including account/bill and service fault/emergency. The call center operates 24/7, with average weekday call volumes of approximately 2,000 calls per day. 90% of calls are answered within 15 seconds. Utilities often separate "service faults" (emergency) call handling from "account enquiry" call handling. Service faults call handling is usually dealt with via a centralized call center operating 24 hours a day, 7 days a week, while a separate call center operating during regular business hours normally handles account enquiries. Recognizing the sense of integrating these two functions, in 2001 Yarra Valley Water made the decision to combine these functions into a single "Customer Contact Center" whereby staff working in the Center handle both emergency/fault and account enquiry calls. This decision reduced costs by eliminating double handling of the above-mentioned calls, improving service through developing call center staff experience and expertise across the broader range of Yarra Valley Water services. In addition, costs could be further reduced as a result of economies of scale achieved through integration to one call center.

FLASH (Fault Logging And Service Help) system

As part of the integration, a unique service fault diagnosis service, FLASH, has been developed. This has enabled a faster response time and improved communication with clients. Since the new system has been introduced response times have been reduced, staff turnover has been significantly reduced, and the answering time has been reduced. Technically, FLASH is a Web application distributed via IE4+ using the Xmarc FIRE Rapid Application Development Tool as the mechanism for viewing map graphics. The application incorporates sophisticated searching routines allowing operators to quickly and simply identify the location of problems from either address information or spatially located data from the map. A key feature of FLASH is that it is designed to mirror the natural flow of conversation with a customer during an emergency/fault-related call. The system utilizes drop-down options, which the operator uses to record the customer's specific problem. In order to facilitate a response, Customer Contact staff can view historical service requests for a property. All current and historical service requests are automatically identified by FLASH for the property at the reported problem location and the surrounding area, allowing operators to immediately identify whether the problem has already been reported and whether the problem location has a key or critical customer, such as a dialysis patient. The system determines the type of fault based on the information provided by the customer and then provides the operator with a script to assist in responding to the customer.

The scripts include such information as when to expect a crew on site, actions the customer can take to minimize any impacts resulting from the fault, e.g. accessing alternative sources of water should supply be interrupted. The system then assigns a priority to the work order. Typical work orders relate to sewer blockages, burst water mains and water quality complaints. These work orders are identified within the GIS thereby enabling Field Maintenance staff and Call Center operators to track the progress of their work and the number of customers affected.

Call Center of the Year Award

Call centers typically have a focus on reducing the Average Handling Time for calls with little requirement for a high level of customer satisfaction. This can result in high staff turnover rates. The emphasis adopted by YVW over the last 3 years was on developing a call center attitude to solving the client's problem at the first point of contact wherever possible. In order to achieve this it was recognized that the call center staff would require a high level of training and morale and hence a low staff turnover rate. The change of the culture of the call center was a long process and involved many small changes that combined to achieve the desired result. The evidence of this change of culture is the presentation of the Victorian Call Center of the Year Award by the Australian Teleservices Association in 2003.

Front Office

YVW has a very small reception area, preferring to serve customers through a range of channels and partners. These partners have service centers distributed across the service area. For example, account payments can be made in person via any post office, and applications for connection and plans can be made through a franchise plumbing supplier, via the "EasyAccess" system, offering 15 minute, automated response and approval for standard connection applications. The EasyAccess system has won International Technology Awards and recognition from the Smithsonian Institution in Washington, DC.

Meter Reading

Meter reading has been outsourced for 10 years. In July 2000 YVW implemented a new "fixed price" contract based on an understanding of the total cost of meter reading, including IT support, back office follow up and other overheads. This made the contractor not only responsible for the collection of meter reading data but the quality of the data and associated rework. The contract includes KPI's to provide customer equity in billing accuracy regardless of the geographical area they live in, but as the contract is "fixed price" there is incentive to drive efficiency to increase profitability of the contract backed up with bonus payments where KPI's are exceeded. In past "rate based" contracts there was little incentive to drive efficiency as they were paid to attempt a read and each subsequent visit was also paid for. This meant all the problems were sent back to the Company to resolve. Since the implementation of the "fixed price" contract, the "meter no read rate" has reduced from 4.9% to 0.56%, all 26 geographical areas (approx 22K meters each) have a no read rate of <1.0% and meters with >3 consecutive no reads has reduced from 1.5% to 0%.

To assist meter readers, YVW is implementing a "virtual office" environment that no longer requires meter readers to attend an office every couple of days. Data is transferred directly to and from the readers HHT via modems installed in the readers homes. YVW is also preparing to trial web based meter reading using mobile phone technology.

Estimated Bill Reduction Program

When a meter reader is unable to obtain reading because access to the water meter is not available, an estimated bill is issued to the customer and a credit or debit adjustment made the next time the meter is read. However, customers dislike estimated bills and are more likely to contact YVW to query an estimated bill or the reason the meter was not read. To improve the situation, the "estimated bill reduction program" was established in 2001/02 with the aim of eliminating estimated bills. To date the program has significantly increased billing accuracy by reducing estimated bills by 72.1% resulting in a reduction of 130,000 estimated bills per annum being sent to customers. The annual estimation rate of 0.96% for 2002/03 also compares favorably with the best performers in the global meter reading industry.

Electronic bill presentment and payment service

YVW has introduced an electronic billing service, which allows customers to receive and pay their bill over the Internet through the Australia Post Internet bill paying service (POSTbillpay). The customer can choose to view only a summary of their bill, or full invoice details. Payments may be scheduled (i.e. programmed to occur at a later, specified, date, provided this is before the due date for the bill), allowing the customer to organize payments for a range of bills on one on-line session. Payments can be arranged from bank accounts or from credit cards. In addition, information usually provided with paper bills (such as information on YVW's other services) is provided electronically with the electronic bill.

The service was launched in January 2001 and provides customers with a paperless transaction process, allowing payment at any time of the day from the customer's own home. YVW was the first utility company to provide payment through POSTbillpay.

However the customer take-up of electronic accounts continues to be relatively low (around 1 - 2%) and one of the key issues is the process by which the level of customer take-up can be increased. Greater focus on marketing strategies including customer market research to increase take-up will be implemented over the coming 12 - 18 months.

Other Payment Options

YVW's payment channel strategy has historically focused on customer satisfaction with the secondary goal of efficiency. Consequently the company now has a diverse range of payment channels with multiple channels offering the same service offering to customers (e.g. a customer can pay by phone using a credit card via the IVR payment number, Australia Post and BPay). In the WSAA benchmarking survey YVW was identified as high in terms of customer service due to the large number of channels offered. Over the past two years, over the counter cash payments have decreased by 11%, but remain the most popular payment instrument at 34% of total payments. Credit card payment volumes have grown by 7% and now represent 25% of the total payment volume, which is the second most popular payment instrument. Debit Cards are the fourth most popular payment instrument with their usage growing by 41% to 13% of total payments.

A detailed payment channel review was completed to identify the opportunities to reduce payment channel costs without significantly impacting customer satisfaction. Payment channels offered include check, direct debit, direct credit, credit card, debit card, electronic funds transfer, and cash. The review concluded that YVW's approach should be guided by the following principles:

- YVW has limited influence in customer trends that are driven by external factors and therefore resources are better directed towards increasing the efficiency of growing channels,
- For channels that are growing or constantly popular, the strategy should be to focus on efficiency and if the transaction cost is relatively low, promotion of that channel.
- For channels that are decreasing in popularity, the strategy should be to essentially do nothing.

Services for the visually impaired

YVW now offers their visually impaired customers the option of a large print bill, or one printed in Braille. The concept was developed with the assistance of Vision Australia, and the bill format (both large print and Braille) has been specifically designed for the visually impaired. A trial of the services commenced in January 2001, and YVW believes there are approximately 3,000 visually impaired customers who may benefit from this initiative. YVW intends to promote the service through Vision Australia.

The service provides visually impaired customers with direct access to information about their water service, rather than relying on someone else to read the account for them.

Credit and Collections

YVW currently outsource the majority of their credit management functions to an external debt collection agency. The reason for the contracting out of debt collection was the inability to contact sufficient debtors when compared to the flexibility of resources offered by an outside agency.

The debt collection focus of the contractor is both efficiency in terms of expected recovery rates and customer service with a number of KPI's established for a range of key measures. It differs from the standard debt collection contracts as it focuses more on customer service with key measures around complaints management and the support and referral of customers in financial hardship.

Over the past three years YVW has achieved significant reduction in long-term debtors whilst costs have been reduced through a focus on continuous improvement through process re-engineering and leveraging of the experience of our external debt collection agency.

All customers are placed into designated arrears paths based on their customer type (e.g. residential, concessional, pensioner, Key Account etc), and different credit management approaches are conducted based on the type of customer segment.

Services to Support Customers in Financial Difficulty

Typically when customers suffer financial hardship they ignore mail and phone calls in relation to overdue payments. The end result can be an escalating problem, and unless carefully managed, become an adversarial relationship between YVW and the customer. YVW had two aims with this initiative, to engage the customer and to improve the relationship with them which is likely to, improve the financial recovery of debts. YVW has a number of existing customer support programs. In 1999 YVW developed a Hardship Policy and in 2000 established a Customer Support Team to deal directly with these customers. The hardship program was developed in collaboration with local welfare agencies. The program was based on an expert understanding of why people have problems with bills and developing an approach using financial counseling and incentives.

YVW now takes the attitude that there is no point chasing people for money they do not have. YVW pays for the counseling services, and has a support team of 6 people to assist customers in this regard. Payment plans are based on what the customer can afford, not the level of the debt. If the customer makes 5 consecutive payments under their plan, the 6th is waived. If the customer is still paying 90 days later, the rest of the debt is extinguished. This program won a Prime Minister's award. It has proven to be cost effective as well as the "right thing to do". All members of the team undergo ongoing extensive training provided by the Kildonan Child & Family Services. In addition Kildonan participates with YVW in focus groups and provides assistance in strategic direction and marketing including development of brochures.

In July 2002 YVW commenced site visits as an attempt to support customers who need help via face-to-face contact. The team members make personal contact with the customer and, where appropriate, offer free water audits and installation of water efficient appliances (which have resulted in an estimated 20% reduction in the water bills for the customers). An expansion of the existing Arrange and Save program now means that if the customer meets payments over a 6-month period their old debt is written off. As of June 2003 payment compliance has risen from 77% to 83 %, and payments received from customers experiencing difficulty was up 13 %. In addition YVW has received positive feedback from both financial counseling organizations and customers and has been requested to present their approach to various other service providers.

Arrange & Save Program

This is a loyalty program aimed at encouraging and supporting customers to pay their water account. Arrange & Save rewards customers who meet their agreed payment plan, by waiving a payment after five consecutive payments have been made. This program includes the write-off of long-term debt if the customer continues to meet their repayment amounts over an additional period of approximately six months. In 2003-04, under the Arrange & Save Program, we granted 4,480 credits to customers for a total value of $72,922 and wrote off $237,775 from 1,592 customer accounts.

Meter Maintenance and Related Field Services

Meter services has been contracted out for 10 years. Initially this involved many separate contracts involved in the meter life cycle. i.e. supply, installation, testing, replacement, repair. However, these contracts have been aligned and consolidated into a "meter fleet" management contract. Under this contract the Contractor runs the day-to-day operations in the most cost effective way based on the Companies policies and strategic direction and a schedule of approved rates.

As the contractor provides similar services to other Companies, YVW benefits from economies of scale generated through the contractor's bulk purchasing agreements with suppliers. For new connections, customers (plumbers, developers) make application through a franchised plumbing store (no need to come to a YVW office) the application is processed on line, fees are paid and a work order is electronically logged with the Contractor to install the meter.

On completion, the work order is closed, with new meter data, and the file is electronically interfaced with the billing system so that all new meters are loaded within 24 hours of the meter being installed in the field. Closure of the work order also automatically generates an invoice based on the schedule of rates that is sent to the contract manager for electronic authorization.

The field service activities now form part of the "fixed price" meter reading contract and are undertaken by the meter reading contractor. As such the contractor has greater ownership of data accuracy as rechecks and audits add unnecessary cost to the contract and highlight the importance of first visit data collection.

Innovative and Noteworthy Practices

If a residential customer occupies a property and Yarra Valley Water Ltd fails to meet the service commitments in its customer contract then the customer will receive a reduction in the price for the supply of water or sewerage services payable (price rebate) automatically or upon application.

Complaints Management

Yarra Valley Water views complaints as a very important area of customer feedback and recognizes that complaints present an opportunity to identify areas for business improvement. Complaints or contacts received by YVW via the Energy and Water Ombudsman of Victoria (EWOV) are processed with significant rigor in terms of case history and documentation. YVW have established an "EWOV Complaints Analysis Committee" to analyze in detail each case processed via the EWOV. The Committee meets on a monthly basis to discuss cases in detail, and comprises representatives and senior managers from across the various areas of the business that are involved in service delivery and policy setting. The Committee is chaired by the General Manager Customer Operations. The key objective of the Committee is to identify root-causes of complaints and options for improving service accordingly. A number of improvements have been generated as a consequence of this strategic approach to managing complaints.

Customer Focused Research

YVW conducted a willingness to pay survey, based on service standards, as the basis for its price submission to the ESC, which provided insights into customer views about risk and service improvement. YVW's bill redesign was driven by unproductive customer service calls and was based on focus groups of customers and service staff. Call data was analyzed before and after the redesign, showing a 3.5% reduction in call volumes and a reduction in misdirected payments. The auditors consider this a response to a problem area within the business and part of continuous business improvement.

Six Sigma

YVW identified a need for a more rigorous systematic tool for improving business processes. The Six Sigma methodology has been in use for 20 years, predominantly in manufacturing businesses. YVW is the first water utility in Australia to adopt it and only the second government organization (after State Rail Queensland). The methodology and tools of Six Sigma require the support of the senior management team and intensive training of the implementing staff.

Six Sigma has a strong foundation built of statistical techniques that ensures decisions are based on reliable factual data rather then instinct or opinions. They define processes with failure modes and defects and can check whether variations are due to real information or the measurement system. The example provided of change of ownership notification processing demonstrates improved customer service as well as cost savings. YVW has already recouped the significant cost of training ($220k) with the savings identified in the first few projects. The Auditors consider this initiative to be a worthy innovation through its application of a methodology developed in a quite different business context and the demonstrated achievement of targeted business improvements.

SPLASH Program

Customer service training has traditionally been skills based and only delivered to frontline customer service staff. YVW identified the benefits of training all staff to improve delivery of service. The SPLASH training program differed from other customer service training by: Involving the whole organization not just those in daily contact with external Customers Involving partner organizations Focusing on vision, attitude and behavioral awareness rather than particular skills (dealing with conflict) and functional performance (time) Using a broader range of training techniques to get staff involved (multi media, role playing, interactive) Although the training has recently been completed and the action plan cycle is still in its early stages, monitoring is showing improved customer satisfaction and reduced costs (efficiency improvements, reduced volume).

ABBREVIATIONS

AA	automated attendant
ACD	automatic call distributor
ACH	automatic clearinghouse
ACSI	American Customer Satisfaction Index
AMR	automatic meter reading
A/P	accounts payable
ASA	average speed of answer
ASR	automated speech recognition
ATM	automatic teller machine
AWWA	American Water Works Association
AwwaRF	Awwa Research Foundation
CIO	chief information officer
CIS	customer information system
CIMS	customer information management system
CRM	customer relationship management
CSR	customer service representative
CTI	computer-telephony integration
EBPP	electronic bill presentment and payment
EDI	electronic data interchange
ENDRO	equivalent number of days' revenue outstanding
ESC	Essential Services Commission
FCR	first call resolution
FTE	full time equivalent worker
GPS	geographical positioning system
HHT	hand held terminal
KPI	key performance indicator
I/C	industrial/commercial
IVR	interactive voice response
KM	knowledge management
LAN	local area network
LOC	line of credit
O&M	operations and maintenance

PAC	Project Advisory Committee
PLC	power line carrier
PUC	public utilities commission
QA	quality assurance
QC	quality control
ROI	return on investment
SPI	single page invoicing
SSN	social security number
VRU	voice response unit
WEF	Water Environment Federation
WMS	work management system
WSAA	Water Services Association of Australia

IWA Publishing's authorised EU representative for General Product Safety Regulations is Diane D'Arras, 15 rue Duret, 75116 Paris, France, e-mail: safety@iwap.co.uk.

Printed and bound by CPI Group (UK) Ltd, Croydon, CR0 4YY
17/09/2026
02218804-0017